Climate Mitigation and Adaptation in China

Policy, Technology and Market

"A superb book."

—Ngaire Woods, *Oxford University, Fellow of the British Academy of Social Sciences*

"*Climate Mitigation and Adaptation in China* is a sophisticated study of China's efforts to decarbonize its economy, which is especially valuable for its sensitivity to global policy discussions and its awareness of the close connections between international commitments and domestic public policy choices."

—Robert O. Keohane, *Princeton University, Member of the US National Academy of Sciences*

"The role of China in tackling climate change will be critical to the success in facing up to the world's greatest challenge. This book explores how policy, technology and markets need to work together to create the conditions to move to a green economy as a model for others to follow."

—Philip Allmendinger, *University of Cambridge, Fellow of the British Academy of Social Sciences*

"This is a deep, up-to-date and rigorous analysis of China's strategic, economic, and technological pathways towards carbon neutrality... An essential reading to understand the new role of China in the global governance of climate change."

—Alberto Quadrio-Curzio, *President, the International Balzan Prize Foundation, and President Emeritus, the National Academy of Lincei, Italy*

"This book provides important insights into Chinese thinking and Chinese policy toward decarbonization. It will be of interest to anyone concerned about Chinese policies toward climate change—and, more broadly, to anyone concerned about the potential for successful action to mitigate climate change globally."

—Jeffry Frieden, *Harvard University, Member of the American Academy of Arts and Sciences*

"As a major and growing economy, China's role in meeting a global climate challenge is simply crucial. Professor Fu Jun and his colleagues have done a masterful job of laying out the parameters, policies and prospects of China's path to sustainability. It should be essential reading, domestically and internationally."

—A. Michael Spence, *Stanford University, Nobel Memorial Prize Laureate in Economic Sciences*

Jun Fu · Dongxiao Zhang · Ming Lei
Editors

Climate Mitigation and Adaptation in China

Policy, Technology and Market

冶金工业出版社
Metallurgical Industry Press

Springer

Editors
Jun Fu
National School of Development
Peking University
Beijing, China

Dongxiao Zhang
College of Engineering
Peking University
Beijing, China

Ming Lei
Guanghua School of Management
Peking University
Beijing, China

ISBN 978-981-16-4309-5 ISBN 978-981-16-4310-1 (eBook)
https://doi.org/10.1007/978-981-16-4310-1

Jointly published with Metallurgical Industry Press, Beijing, China
The print edition is not for sale in China (Mainland). Customers from China (Mainland) please order the print book from: Metallurgical Industry Press

This Springer imprint is published by the registered company Springer Nature Singapore Pte Ltd.
The registered company address is: 152 Beach Road, #21-01/04 Gateway East, Singapore 189721, Singapore

Preface

Climate change is a huge challenge facing humanity in the twenty-first century. The 2015 Paris Agreement represents a global consensus to keep global temperatures well below 2°C above the pre-industrial levels and to pursue further efforts to limit the temperature increase within 1.5°C. In 2018, the UN Intergovernmental Panel on Climate Change (IPCC) has further warned that the impact of above 1.5°C of global warming will spell disasters to the ecosystem on Earth. As a response to the call for renewed efforts at mitigation and adaptation to climate change, in 2020, China has again pledged to the international community that it will peak CO_2 (carbon dioxide) emissions prior to 2030 and try to achieve carbon neutrality by 2060—indeed a tall order to fulfill given China's current stage of development.

Fight against climate change calls for concerted efforts by all walks of life, both domestic and international (e.g., if every individual in China were to reduce one ton of CO_2 each year, it would amount to a total reduction of 1.4 billion tons of CO_2, which is roughly an equivalent of the yearly CO_2 emissions by the Chinese steel or cement industry). Similarly, as reflected by the on-going research on climate change at Peking University, cross-disciplinary approach is highly desirable. Presented here is a book-length study organized through what one may call a triple helix of three schools at Peking University—the National School of Development, the College of Engineering, and the Guanghua School of Management. The scholars and experts associated with these three schools cover three domains of knowledge—policy, technology, and markets—that are essential in fight against climate change.

China is a large developing country with 1.4 billion people in the midst of industrialization and urbanization. GDP per capita in China in 1979 is about US$156; it is about US$ 10,000 in 2020; urbanization is 18% in 1979; it is close to 60% in 2020. Topping output of 220 of 500 major industrial products in the world in 2011, China became the world's largest industrial country. As a consequence, currently China is the largest CO_2 emitter in the world, and it contributes roughly a quarter of global CO_2 emissions, although on per capita basis, China has remained moderate—about 7 tCO_2 per year, as compared to 16 tCO_2 for the U.S. and 5 tCO_2 for the world average.

Take the iron/steel sector in China for instance. The sector has a total of some 500 firms with a total output of crude steel reaching 1.065 billion tons in 2020, accounting for over 57% of the world's total. In terms of CO_2 emissions, it accounts for over 60% of total CO_2 emissions from global steel production and about 15% of total domestic industrial CO_2 emissions, following the power (including heat) sector (about 50%) and above the cement sector (about 12%).

But China has been working hard at mitigation and adaptation to climate change and trying to de-couple economic growth from CO_2 emissions at the earliest date possible. And the present study provides a prism through which to see China's efforts and appreciate its magnitude. The study is organized around the three linchpins, i.e., Part I, Policy; Part II, Technology; and Part III, Market. This study ends with a projection to 2060 in Part IV, by which China has pledged to achieve carbon neutrality. China has to be an integral part of the global efforts in fight against climate change—the most consequential threat facing humanity. As the latest UN Intergovernmental Panel on Climate Change suggests, without concerted efforts, the dire situation will only get worse. Everyone must act now—with a sense of urgency—to address the crisis.

What follows is a synopsis of the these principal building blocks—as well as the overall research design—of the study.

Part I Policy

At a fundamental level, climate change reflects the logic of negative externalities of what scholars have described as "the tragedy of the commons." In order to prevent that tragedy from happening, public policy becomes indispensable. To understand China's evolving policy framework to decarbonize its economy, a principal theme to watch is the dynamic interaction between China's international commitments and domestic agenda setting with respect to climate change. At the risk of oversimplification, three sets of milestone international commitments can be identified for China's efforts of mitigation and adaption to climate change.

Under the principle of "common and differentiated responsibilities" of the UNFCCC (UN Framework Convention for Climate Change), in 2009, China pledged to the international community to reduce carbon intensity per unit of GDP by 40–45% by 2020 from the level of 2015. In 2015, China pledged again to peak carbon emissions around 2030 and reduce carbon intensity by 60–65% by 2030 from the level of 2005. In 2020, building upon previous successes, China again pledged that it would peak carbon emissions prior to 2030 and try to achieve carbon neutrality by 2060. This pledge was particularly significant. Indeed, shifting from carbon *equity* toward carbon *neutrality* indicates a change of political philosophy from particular justice to general justice. (Note that China, about 157 tCO_2, was still below the world average, about 209 tCO_2, in terms of per capita accumulated CO_2 emissions from 1900 to 2019. By comparison, it was about 1218 tCO_2 for the U.S., 930 tCO_2 for the U.K., and 516 tCO_2 for France). These international commitments, in turn, have been successively factored into China's 11th Five-Year Plan (FYP 2006–2010), 12th FYP 2011–2015, 13th FYP 2016–2020, and 14th FYP 2021–2025, and together they have constituted an evolving policy framework, shaping the country's efforts to decarbonize its developmental path.

Initial evidence is compelling that China has taken these commitments seriously. In 2019, for instance, China has fulfilled its 2009 pledge one year ahead of schedule, with carbon intensity reduced by 48.1% from the 2005 level and non-fossil fuels raised to 15.3% as a share of energy structure. In 2016–2019, a total of 1.4 billion tons of CO_2 emissions were reduced due to work at energy efficiency nation-wide. As evidenced in the steel sector, with both "sticks" and "carrots," the Chinese government had aggressively pursued a low-carbon strategy, by way of reducing inefficient overcapacity and enhancing energy efficiency in the 13th FYP (2016–2020). And the steel sector has been planning for a super-low-carbon strategy for the 14th FYP (2021–2025). Arguably, the overall current approach seems to be more fiat than market. But the trend is increasingly market-oriented. Indeed, having gained experience from its eight (7+1) regional ETS pilots, China operationally rolled out a national carbon market (or ETS) in 2021, which promised to become the largest in the world.

The Paris Agreement set out a global action plan to keep the world on track by limiting global warming to well below 2°C above pre-industrial levels in the long-term and urged contracting states to undertake rapid carbon emission reductions with best available technologies (BATs). As a contracting state to the Paris Agreement, China has been keen on BATs, as illustrated by its policies to encourage and shape the development of carbon capture, utilization, and storage (CCUS). Indeed, as a global fieldtrip would attest, in the last few years China has been catching up rapidly in this field and has become one of six leading countries (the USA, Canada, UK, Norway, Japan, China) in the world. Thus far, whereas most

CCUS pilots and demonstrations in China are in the power sector (accounting for about 50% of total industrial CO_2 emissions), its steel sector (accounting for about 15% to total industrial CO_2 emissions) seems increasingly posed to set up CCUS pilots as well in the 14th FYP (2021–2025).

Part II Technology

Generally speaking, there are three broad categories of technologies of decarbonization, i.e., low carbon, zero carbon, and negative carbon. To illustrate, hydraulic, wind, and solar power are examples of zero-carbon technologies, but the caveat here is the supply incongruence or instability of wind and solar energy for the grid, unless battery technology has made hitherto unknown breakthroughs. Currently, however, only about 15% of China's grid capacity is VRE (variable renewable energy) compatible for wind and solar power, the sources of which vary across different regions and are unstable by nature, although their cost, excluding storage, has come down dramatically in recent years. The challenge here is for the grid to move up to higher phases of VRE integration. Or an alternative strategy is to develop green (as opposed to grey) hydrogen. That is, to use wind or solar power to produce green hydrogen locally and then store and transport it cross-regionally. In this case, attendant technology for its storage and transport must also be developed on scale. Note as well, on the endless frontier of science to watch, nuclear fusion of deuterium and tritium, two hydrogen isotopes, if successful, would mean the ultimate clean energy that is almost boundless.

Various techniques or methods of energy conservation and energy efficiency—aggressively pursued in China's iron/steel sector thus far—typically embody low carbon technologies. By contrast, carbon capture, utilization, and storage (CCUS), including both onshore and offshore sequestration, represents negative carbon technology. Looking forward, a strategic pathway for China to move toward carbon neutrality by 2060 is likely to feature a dynamic mix in different proportions of all three categories of technologies.

To cut into the web of green technologies, our research strategy is to address the most complex and still evolving one head on, i.e., CCUS, believing that, having done the most difficult, the rest—e.g., hydraulic, wind, and solar power, etc.— would be relatively easier to piece together. After all, much research has already been done on the latter part of the picture, the basic contour is clear, and we do not have to start from scratch. Accordingly, lending credence to the strategic roadmap based on simulation for the future deployment of CCUS in the Chinese steel sector in the previous chapter, Part II of this study takes an in-depth look at CCUS as an emerging strategic reserve technology on industrial scale for direct abatement of CO_2 emissions. Under country-specific engineering conditions, the study begins by methodically going through a typical iron/steel plant in China. In the process, it provides a systematic analysis of the appropriate options as well as potential pitfalls of CO_2 capture technologies—physical or chemical or both—in steel-making processes, including options for CO_2 transport and storage. In addition, cost estimates, together with tech-performance, of a full-chain CCUS are also provided to assist planners or decision-makers in identifying most cost-effective pathways.

For example, integrated infrastructural outlays involving optimal CO_2 source–sink matching can presumably reduce unit cost of a CCUS-retrofitted plant operating in a hub-spoke cluster. In light of the source–sink matching and the maturity of various CCUS technologies for CO_2 emission reduction in the near, medium, and long term, there is evidence to suggest that the steel sector should adopt the strategy linking enhanced oil recovery (EOR) and enhanced coal bed methane recovery (ECBM) technologies as the near- and medium-term utilization and storage solution and storage in deep saline aquifers as the long-term storage solution. At present, EOR and ECBM in China, experimentally or otherwise, account for only a few million tons of CO_2 used annually–indeed a tiny part of the 10.4 billion tons of CO_2 that China emitted in 2020. Beyond the steel sector and for considerations of scale economy and public good, it also makes sense for the government to plan, design, and invest in the basic infrastructure of hub-spoke transport clusters to facilitate or speed up the deployment of integrated full-chain CCUS projects.

Currently, however, while slow progress on CCUS pilots or demonstrations has been under way in China and overseas, huge initial investment and lack of viable business models for cost considerations are seen as the main obstacle to rapid development of CCUS, globally. The International Energy Agency (IEA) has warned that CCUS is not on track for Paris Agreement targets and picking up the pace of CCUS deployment is urgent. In China, currently CCUS pilots are mostly in coal-fired power plant and none has captured CO_2 emissions in the steel and the cement sector. But as China revs up efforts to achieve carbon neutrality by 2060, it is expected that the deployment of CCUS would quicken steps in fossil fuels-intensive industries, including in the iron/steel sector. Indeed, if the rapid unfolding of the fast-speed railway network in recent years is any indication, China would also be particularly good at building multiple CO_2 source–sink clusters for CCUS to scale up. And one would anticipate that work is likely to start during the 15th FYP (2026–2030), if not earlier. And if it does, it would be better to pilot first rather than big-bang approach, because CCUS has remained complex, costly, and uncertain in the nexus of all green technologies thus far. As such, in strategic sequencing, it makes sense to see CCUS as a reserve technology or "the lender of last resort", so to speak, to borrow a term from finance.

Part III Market

Right sequencing is a critical component of a successful strategy—a lesson that China has learned from its own experience of reforms and opening-up in the past 40 years. This logic seems to have applied to the development of carbon market in China as well. That is, "testing it out before scaling it up," so to speak. On the heels of the previous part on technology, lest one forget, markets as humanly-made institutional arrangements are institutional technologies, the design of which has to deal with both politics and economics and as such can either speed up or slow down the rate of production of ideas, products and services. Fundamentally, markets and the rule of law, as is often said, are the two sides of the same coin. In building carbon market, in 2013, after two years of preparations, China rolled out seven regional ETS (emission-trading system) pilots (i.e., in Beijing, Chongqing, Guangdong, Hubei, Shanghai, Shenzhen, and Tianjin) as a first step. The vision was—if these regional ETS pilots were successful—to build a unified national ETS starting in 2017, which eventually would become the largest ETS in the world, covering well over 6 billion tons of CO_2 emissions, or more than half of China's total CO_2 emissions.

Theoretically speaking, ETS built on the theme of "cap and trade" can reduce CO_2 emissions at a lower cost than direct government regulations such as mandated technologies and performance standards. Thus far, the theoretical proposition has been validated by the experience in the USA and the European Union. Are the regional ETS pilots really working in the Chinese context of regional disparities? To assess their validity in terms of behavioral changes on the part of polluting firms toward de-carbonization is thus the focus of Part III. For the sake of analytical rigor, the present study has made efforts to go beyond mere descriptive statistics by deploying the method of synthetic control in evaluating the efficacy of these regional ETS pilots.

Going beyond a prima facie case, the main findings of the investigation are as follows. First, even at this early stage, except Tianjin and Chongqing, there is evidence of varying degrees that these regional ETS pilots have contributed to the reduction of CO_2 emissions, with Hubei ETS pilot being the most salient case. Second, the early success is nevertheless uneven across different regions in China, suggesting that caution is needed when considering scaling up regional

ETS pilots to a unified national carbon market after 2017. Given the varying industrial structures in different regions across China for historical reasons, strategically a gradual rather than a big bang approach—with different phase-in periods for various sectors—to the construction of the national ETS would seem appropriate. Or, one could run the risk of making rich regions richer and poor regions poorer.

Third, as information and transparency is essential for market operations, a robust MRV system (monitoring, reporting, and verification) is critical, as China moves further ahead with the construction of the unified national carbon market, as well as with green financing. According to the People's Bank of China (PBOC), by 2030 China will need to invest about RMB 2.2 trillion on a yearly basis to decarbonize its economy, and the amount will further increase to about RMB 3.9 trillion from 2030 to 2060. To meet the need, both public and private funds must be mobilized.

By 2020, all of China's regional ETS pilots had covered a total of some 3000 firms and accumulated a CO_2 trade volume of some 430 million tons, or about RMB 10 billion. Yet the current carbon price of these regional ETS pilots, averaging about RMB 30/tCO_2, seemed too low to spur serious strategic shifts in terms of investment in new technologies such as CCUS. In principle, carbon price must reflect environmental cost, and, here, China still has a long way to go. By international comparison, carbon price in the USA and EU was about US$ 15–16/ton in 2018. Globally, unless carbon price increases dramatically in the years to come, say over US$ 70/tCO_2, to galvanize a more robust path to decarbonization, it would be difficult to achieve the goal of the 2015 Paris Agreement, i.e., keeping global temperatures well below 2 °C above the pre-industrial levels.

In this connection, Part III also explores various design options of ETS on how best to incentivize CCUS from the perspective of a business model, for investment decisions with respect to green technologies (not necessarily CCUS) are ultimately made at the firm level, and sensitive to public policy and institutional environment. Take it as a brain teaser to think further ahead, now that China has rolled out a national ETS in 2021, bringing the world's largest ETS online after three years of preparation on the basis of its experience of eight regional ETS pilots. As a first step, the national ETS now regulates some 2,200 firms from the power sector (including combined heat and power, as well as captive power plants of other sectors), and its scope is expected to expand further in the future, to cover refining and petrochemicals, chemicals, building materials, steel, nonferrous metals, paper, and aviation as well. If and when the national ETS expands to cover all eight sectors, it will cover well over 60% of China's total emissions down the road.

Part IV Projection

Taking stock and looking forward, in 2020, China has pledged to the world that it would peak carbon emissions prior to 2030 and try to achieve carbon neutrality by 2060. The task is daunting. Put in international perspective, China's deadline is very tight. Whereas most advanced economies have planned to spend 50–70 years from carbon peak to carbon neutrality, China has only given itself a 30-year transition period. Thus, in the years ahead, China must move aggressively to decarbonize its economy, and in the process, transitional risks can be enormous, especially for resource-intensive and fossil fuel-intensive industries. In 2020, China emitted about 10.4 billion tCO_2—or about 7.4 tCO_2/per capita—about 92% of which comes from the burning of coal (about 7.4 billion tCO_2), oil (1.5 billion tCO_2), and natural gas (0.6 billion tCO_2). As can be imagined, the challenge for China to de-carbonize its economy within the next 40 years can be mind-boggling.

Be that as it may, there are basically four general approaches for China trying to meet the challenge lying ahead, i.e., public policy approach (PPA), market-based approach (MBA), carbon capture and storage (CCS), and nature-based solutions (NBS), such as forest sink. Based the conceptual framework of PPA+MBA+CCS+NBS (or PMCN model for short), a big part of Part IV represents efforts to project forward by way of disciplined exercises of simulation of various scenarios, such that one can see more clearly what lies ahead and reckon how best to respond. On the basis of these simulations or projections, the main takeaways are as follows:

First, as was true during the so-called new normal period, when growth rate of Chinese heavy industry (i.e., energy, cement, steel, automobiles, shipbuilding, construction, etc.) dropped dramatically from above 10% in 2008–2011 to 6% in 2015, generally economic slowdowns have lessened the steep upward trend of CO_2 emissions, thus helping China peak CO_2 emissions by 2030. Conversely, the simulation exercises also show that all the strategies of carbon emission reduction, including ETS, come at the cost of GDP loss, with modest declines in employment, investment, and consumption. That said, a better calibrated mix of policy instruments will take smaller costs of economic loss. In this context, in the various PPA+MBA combinations, high carbon price rather than low energy enhancement is a preferred option. In longer terms, market approach is a preferred choice over administrative fiat.

Second, to peak CO_2 emissions before 2030 and achieve carbon neutrality by 2060, along the spectrum of PPA+MBA+CCS+NBS, all the instruments currently available (e.g., improvement in energy efficiency, cost reduction in renewable energy, and carbon pricing) are all necessary, but not enough even in a full combination of all above instruments. New instruments, especially for a more powerful shift from fossil to no-fossil fuels, are needed to achieve the carbon neutral goal in time. Here, carbon tax and green finance, including green loans, green bonds, and green projects, are important areas that need further explorations. Lying between PPA and MBA, carbon tax and green finance could be a potent tool to facilitate technology shifting or even leapfrogging from fossil to non-fossil fuels. And just as in ETS, to develop a robust, sophisticated and standardized MRV system, not just domestically but also internationally, including good metrics or protocols of ESG to avoid "green washing" or reduce moral hazards, is also critical in green finance, if only for the sake of risk assessment, e.g., green assets vis-à-vis brown assets. The issue of moral hazards would arise, where people's interest and risk are not well aligned so they may have incentives to take on greater risk than they would if no one else effectively contributed insurance. Here, we must encourage international cooperation and create the right incentive structures, with respect to risks, rewards, and responsibilities.

Finally, combined rather than separate PPA aiming at de-carbonization will reshape the future structure of energy demand most powerfully. Specifically, renewable energy will be on the rise, whereas fossil fuels, especially coal, will decrease. Yet a caveat is in order. Even in the most comprehensive of all PPA scenarios available today, by 2060, renewable energy will only account for less than 80% of energy structure, whereas fossil fuels still account for more than 20%. In this light, introducing carbon tax and promoting the electric vehicle sector and other forms of electrification are important for further reductions in fossil energy, but the prerequisite is that the sources of electricity have to be clean (Currently, however, fossil-fuels power plants account for roughly 70%, the rest is non-fossil—hydraulic roughly 15%, wind 7%, nuclear 5%, and solar 3%). And hopefully, variables exogenous to the current parameters of the simulation model—e.g., dramatic expansions of green finance, technology breakthroughs, including additional PPA instruments—may come along the way, thus increasing the chance that China achieve carbon neutrality in time.

Admittedly, accurate and trustworthy predictions are difficult, even when people do their best to model everything relevant. The inputs and assumptions that enter any particular model might significantly affect a conclusion. It is therefore critical to be straightforward about uncertainties if a prediction is to have any value. Note as well, there is a difference between simulation and emulation. Whereas the former is basically linear in thinking, the latter is characterized by both deductive and inductive tinkering, constantly searching for order out of complexity and even chaos. Fundamentally humans are good at emulation. Indeed, facing uncertainties and unknowns, one has to stand ready to adjust and adapt as circumstances change. Accordingly, as a research strategy for such a complex issue as climate crisis threatening humanity, one ought to conceptualize like a "philosophy-king", straddling different disciplines of knowledge for a broad and yet logically consistent vision, and act like an academic "foot soldier", who is well armed with the "state-of-the-art" tools or techniques of rigorous analysis and investigation. After all, philosophical views could affect the conceptual framework one deploys to describe predictions, even though not the predictions themselves. And indeed, it is at least in this spirit that the current book is orchestrated and written, leaving ample room for adding details and making corrections down the road, as contingency preparedness by way of both deductive and inductive tinkering increases resilience in transition. Our hope is to provide a rough map rather than the actual driving itself, as we move towards the middle of the 21st century.

To sum up, to peak carbon emissions prior to 2030 and achieve carbon neutrality by 2060 is a very tall order, fraught with uncertainties and unknowns down the road. What is certain is that China has to undergo a veritable green revolution —from what the Chinese would say "*Gongye Wenming*" (industrial civilization) to "*Shengtai Wenming*" (eco-civilization), which, broadly speaking, would also include protecting biodiversity and healing geo-environmental degradation – both are relevant topics but nevertheless beyond the scope of this study. In the process, as the world is intrinsically interconnected, none of us will be spared the harsh realities of ecological crisis, which means we are all in this together—"*Tongzhou Gongji*" (all in the same boat).

Beijing, China Jun Fu

Acknowledgements

Initial research included in this book was financially supported by BHP through a project-based grant to Peking University which began in 2017 through 2020 with a focus on carbon capture, utilization and storage (CCUS) in the steel sector. Follow-up research and editing of the book in 2021 through its publication was financially supported by the Leo Koguan Foundation and the Deloitte Touche Tohmatsu CPA Ltd. through their educational and CSR programs. On behalf of our colleagues involved in this cross-disciplinary study, the editors of this book—who have led the research teams on climate change from the National School of Development, the College of Engineering, and the Guanghua School of Management at Peking University—gratefully acknowledge the generous support.

For logistic support, we would like to thank Peking University Public Policy Forum International, which has hosted many stimulating workshops and seminars on climate change, among others. We are also grateful to Yin Jianhong, Zhao Ting, and Hu Weiyi for administrative support.

Contents

About the Editors

Jun Fu, Esq., is Professor of Political Economy and Public Policy at the National School of Development, Peking University. He received his Master's degree and Ph.D. from Harvard University and is the first Chinese to have been elected Foreign Academician of the Bologna Academy of Sciences in its long history. He is the author of *Institutions and Investments* (Studies in International Economics, the University of Michigan Press), and *The Dao of the Wealth of Nations* (Peking University Press). Previously, he held faculty positions at the School of Economics and Management, the School of Public Policy and Management, Tsinghua University, and the School of Government, Peking University. He has served, among others, as Member of the Listing Committee of Shenzhen Stock Exchange, Vice Chair of the World Economic Forum's Global Agenda Council on New Growth Models, and Advisor to Chairman of the Executive Council of UNESCO. As a globally known thought leader, he was twice featured in the *Outlook on the Global Agenda* (World Economic Forum). Appointed by its Governing Board, he has also served on Visiting Committee for International and Area Studies across Harvard University.

Dongxiao Zhang is Chair Professor at College of Engineering, Peking University. An internationally well-known expert in unconventional oil and gas production, groundwater hydrology, and geological carbon sequestration, he had held positions as Senior Scientist at Los Alamos National Laboratory, Miller Chair Professor at the Department of Petroleum and Geological Engineering at the University of Oklahoma, Chair Professor at the University of Southern California, and Executive Dean of Graduate School and Dean of College of Engineering at Peking University. He has published over 220 peer-reviewed papers and authored 2 books, *Stochastic Methods for Flow in Porous Media: Coping with Uncertainties*

(Academic Press), and *Theory, Modeling and Field Investigation in Hydrogeology* (Geological Society of America). He earned both his Master's degree and Ph.D. in hydrology and water resources in 1992 and 1993, respectively, from the University of Arizona. He is Member of the U.S. National Academy of Engineering, Honorary Member of Society of Petroleum Engineers, and Fellow of Geological Society of America.

Ming Lei is Full Professor at Guanghua School of Management, Peking University. He is the author of *Green Input-output Accounting: Theory and Application, Green Accounting of China* (1992–2002) and *Integrated Analysis of China's Natural Resources-Economy-Environment* (1992–2002) (Peking University Press, 2000/2010/2011). He was Associate Professor of Donated Chair by KENSAI Power of Japan, Graduated School of Energy Science, Kyoto University, Japan in 1998, and has held numerous honorary and academic positions, including President elect of Asia Pacific DSI, Senior Expert of Chinese National Green GDP Accounting Project, Fellow of International Sustainable Energy Association (ISEA), Fellow of Chinese Environmental Science Society, Fellow of Chinese Operation Research Society, Fellow of Chinese Environmental and Culture Society, Fellow of Chinese National Accounting Society, and Fellow of Chinese Energy Society. A Member of the Editorial Broad of *Economic Science*, he is Dean of the Institute on Poverty Research at Peking University and Honorary Professor of the University of Edinburgh.

Abbreviations

AAUs	Assigned Amount Units
BATs	Best available technologies
BECCS	Bio-Energy CCS
BFG	Blast furnace gas
BFs	Blast furnaces steelmaking
BOF	Basic oxygen furnace
CAMU	China Association of Metal Scrap Utilization
CCER	China certified emission reduction
CCS	Carbon capture and storage
CCUS	Carbon capture, utilization, and storage
CCX	Chicago Climate Exchange
CDM	Clean Development Mechanism
CERs	Certified emission reductions
CNPC	China National Petroleum Corporation
CO_2	Carbon dioxide
CO_2e	Carbon dioxide equivalent
CO_2-EOR	Enhanced oil recovery
CSLF	Carbon Sequestration Leadership Forum
DAC	Direct air capture
DICE	Dynamic integrated model of climate and the economy
DRI	Direct reduce ion
EAF	Electric arc furnace
EC	European Commission
ECBM	Enhanced coal bed methane
EIA	Environmental Impact Assessment
EIB	European Investment Bank

EOR	Enhance oil recovery
ERUs	Emission reduction units
ESA	Electric Swing Adsorption
ETG	Emissions Trading Group
ETS	Emission trading system
EU ETS	EU Emissions Trading System
EU	European Union
FFCER	Fujian Forestry Certified Emission Reduction
FYP	Five-Year Plan
GCCSI	Global Carbon Capture and Storage Institute
GDP	Gross domestic product
GHG	Greenhouse gas
HBFG	Hot blast furnace gas
IEA	International Energy Agency
IET	International Emission Trading
IPCC	Intergovernmental Panel on Climate Change
IUCN	The International Union for the Conservation of Nature
JI	Joint implementation
KPI	Key Performance Indicators
LDG	Linz–Donawitz gas
LSIPs	Large-scale integrated projects
MAF	Electric arc furnace
MBA	Market-based approach
MEA	Mono Ethanol Amine
MEE	Ministry of Ecology and Environment
MEP	Ministry of environmental protection
MIIT	Ministry of Industry and Information Technology
MIRV	Monitoring, Reporting and Verification
MNR	Ministry of Natural Resources
MOF	Ministry of Finance
MOST	Ministry of Science and Technology
MOT	Ministry of Transport
MSPE	Mean Square Percent Error
MTCE	Million tons of coal equivalent
NAP	National Allocation Plans
NBS	Nature-based solutions
NCSC	National Center for Climate Strategy and International Cooperation
NDC	Nationally determined contributions
NDRC	National Development and Reform Commission

NPC	National People's Congress
NRA	National Railway Administration
NSW	National Trust of Australia
NTCC	National Technical Committee on Carbon Management
PHCER	Pu Hui Certified Emission Reduction
PPA	Public Policy Approach
PSA	Pressure Swing Adsorption
RICE	Regional dynamic integrated model of climate and the economy
SAC	Standardization Administration of China
SACC	Strategy Analysis of Climate in China
SFA	State Forestry Administration
SO_2	Sulphur dioxide
TCE	Ton of coal equivalent
TGRBF	Top Gas Recycle of Blast Furnace
TRT	Top Gas Pressure Recovery Turbine
TSA	Temperature Swing Adsorption
UNFCCC	United Nations Framework Convention on Climate Change
VPSA	Vacuum Pressure Swing Adsorption
WTO	World Trade Organization

List of Figures

Chapter 1

Chapter 2

Chapter 3

Chapter 4

Chapter 5

Chapter 6

Chapter 7

List of Tables

Chapter 5

Chapter 6

Chapter 7

Part I
Policy

Mitigation & Adaptation—China's Evolving Policy Framework

1

Jun Fu, Qiang Zhou, Ruqiang Zou, and Dandan Zhang

Climate change caused by anthropogenic greenhouse gas (GHG)[1] emissions from industrialization is a global challenge. The 2015 Paris Agreement represents a global action plan to set the world on track to avoid dangerous climate change by limiting global warming to well below 2 °C above pre-industrial levels in the long-term, and to pursue best efforts to limit the increase in warming to 1.5 °C. To achieve the goal, the Paris Agreement underscored the need for global GHG emissions to peak as soon as possible and for the contracting states to the Agreement to undertake rapid carbon emission reductions with best available technologies (BATs).

[1] Gases that trap heat in the atmosphere are called greenhouse gases. Global GHG emissions due to human activities primarily consist of Carbon dioxide (CO_2), Methane (CH_4), and Nitrous oxide (N_2O). These three types of gases represent 76%, 16% and 6% of total global GHG emissions respectively. In 2017, for example, global GHG emissions was 50.9 gigatons in CO_2 equivalent. See U.S. Environmental Protection Agency and PBL Netherlands Environmental Assessment Agency.

The original version of this chapter was revised: A sentence has been added on page 29. The correction to this chapter can be found at https://doi.org/10.1007/978-981-16-4310-1_8.

J. Fu (✉) · D. Zhang
National School of Development, Peking University, Beijing, China
E-Mail: fujun@pku.edu.cn

Q. Zhou
School of Government, Peking University, Beijing, China

R. Zou
College of Engineering, Peking University, Beijing, China

J. Fu et al. (eds.), *Climate Mitigation and Adaptation in China*,
https://doi.org/10.1007/978-981-16-4310-1_1

Due to its large population (1.4 billion people) and rapid economic growth in the midst of industrialization, currently China is the largest CO_2 emitter in the world, contributing more than a quarter of global carbon emissions. But China has been working hard at mitigation and adaptation to climate change with its increasingly progressive international commitments to reduce CO_2 emissions, which in turn have facilitated its domestic agenda setting for developing a low-carbon and green economy. In the past two decades, there have been three milestone announcements of China's green pledge to the international community. In 2009, during the Copenhagen Climate Change Conference, China pledged to cut CO_2 emissions per unit of GDP by 40–45% by 2020 from the 2005 level—a commitment that China has achieved by the end of 2019.[2] In 2015, during the Paris Climate Change Conference, China again committed to peak carbon emissions around 2030 and reduced carbon intensity of GDP by 60–65% by 2030 from the 2005 level. In 2020, at the General Debate of the 75th Session of the UN General Assembly, Chinese President Xi Jinping pledged again that China will peak its carbon emissions *prior to* [emphasis added] 2030,[3] and become carbon neutral by 2060.

The aim of this chapter is to outline the evolution of China's policy or regulatory framework to de-carbonize its economy. It does so through the prism of the steel sector and with a special attention to the emerging technology of carbon capture, utilization and storage (CCUS). Where relevant, international comparisons are also made so that we have a sense of where China stands now in terms of both policy and technology for green growth.

The rest of this chapter is organized as follows: Section 1 presents a brief overview of China's industrial growth, including the steel sector, in the past 2–3 decades, in terms of growth trends, energy consumption and carbon emissions. Section 2 discusses China's policy framework to decarbonize the industry. In the process, we will identify the key government departments tasked with shaping the country's strategy for sustainable economic and social development. Section 3 puts China in international perspective in connection with CCUS—an emerging technology increasingly recognized as having great potentials for direct abatement of CO_2 emissions (Global CCS Institute, 2015). Section 4 concludes by a brief review on China's CCUS developments in terms of both policy and practice.

---

[2] By the end of 2019, the reduction was 48.1%, and non-fossil fuels as a share of energy structure reached 15.3%.

[3] To achieve this goal, it means that carbon intensity has to go down by 60–65%, non-fossil fuels as a share of energy structure have to go up to 20%; and forest sink has to increase by 4.5 billion square meters from the 2005 level.

1 Section 1

1.1 Chinese Industrial Sectors

Since economic reforms and opening-up initiated in 1978, China has been growing at almost double-digit rates for many years, which was fueled by the fast growth of the industrial sector. By 2011, the China's total output of over 220 industrial products among 500 major ones were in the first place of the world. That would make China the world's largest industrial country. In terms of the industrial structure, the second industry had long been in the dominate position before it was overtaken by the tertiary industry in 2012. During China's 11th Five-Year Plan (FYP) (2006–2010), the ratio of heavy to light industries was 7:3, significantly higher than the current level and even surpass the highest level of many developed countries.[4]

The secondary industry of China has been fueled by the rapid urbanization process in China. The urbanization rate of China was about 18% in the late 1970s, it has reached 64% in 2020. Since the mid-1990s, heavy industries—automobiles, petrochemicals, and heavy-duty equipment, etc. —received massive government support at both the national and the local levels, reflecting China's industrial policy at that time. After the worldwide financial crisis in 2008, however, the Chinese economy has been slowing down, and eventually entered a new period of which Chinese President Xi Jinping called the "New Normal".[5] With the "New Normal", real GDP growth rate of China has dropped dramatically—from an average of 10.76% in 2003–2011 to 7.2% in 2012–2017 (National Bureau of Statistics, 2018). Despite the growth of industrial value added has slow down, the industrial sector has remained its original scale in China. In 2020, the value added of the secondary industry was RMB 38,404 billion, accounting for 37.8% of China's GDP (Fig. 1).

[4] Starting in 2013, the National Bureau of Statistics (NBS) of China no longer uses light industries or heavy industries categorization. Instead, NBS uses a classification system that conforms to ISIC. See http://www.stats.gov.cn/tjsj/zxfb/201402/t20140224_514970.html.

[5] The term "new normal" first came to public discourse during President Xi Jinping's visit to Henan in May 2014. Under the "new normal", the importance of growth speed is eclipsed by immensely complex structural reforms to transform the economy to one that relies more on the services sector, domestic spending and innovation, than manufacturing and exports. http://www.aisixiang.com/zhuanti/175.html.

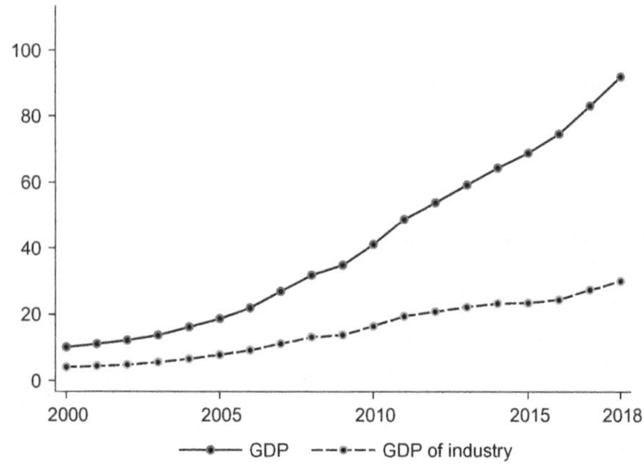

Fig. 1 Output of total economy and industry sector (RMB trillion). *Source* China Statistical Yearbooks

1.2 The Steel Sector

The steel sector plays an important role in the Chinese economy. China has been the world's largest iron and steel producer since 1996. Crude steel production was 1.6 million tons in 1949 (when the People's Republic of China was established), it surged to 804 million tons in 2015; on per capita basis, China reached 586 kg of steel production, about 2.6 times the world's average, next only to South Korea. In the fleet of China's iron and steel plants, huge disparities existed in production capacity, energy-saving technology, and product structure. Yet long-process iron/steel production remains the mainstream, whereas short-process steelmaking accounts for only about 10%.[6] Compared with the United States, the share of electric arc furnaces (EAF)—a short-process steel production—was 68% in 2018.[7] (Fig. 2).

The impact of the "New Normal" was tremendous on the China's industrial sector, in particular on the steel sector—a topic that we will pick up and discuss further in Chapter 2. We emphasize here that the steel sector had confronted severe

[6] There are two processes in iron and steel production: a long process and a short process. The main difference is that the long process takes iron ore and coke as the main raw materials, melts them and then converts them to steel; while the short process uses scrap steel as a raw material to melt steel. More on iron/steel making will be discussed in Chapters 3 and 4 of this study.

[7] World Steel in Figs. 2019, World Steel Association.

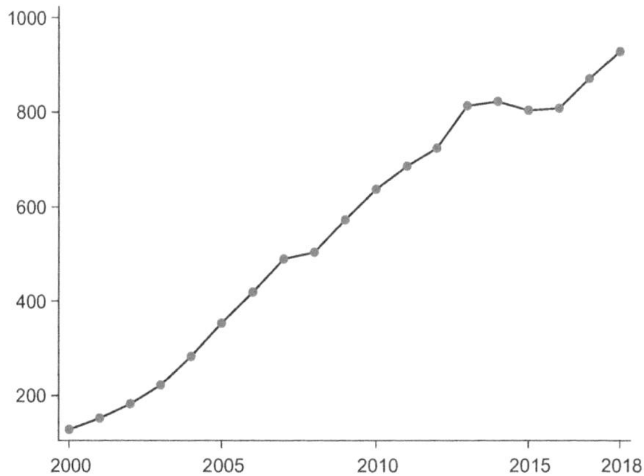

Fig. 2 Output of crude steel (million tons). *Source* Statistical Abstract of China

overcapacity issues. Capacity utilization rate in the steel sector fell from 88.7% in in 2006 to 64.8% in 2015. Since then, like many other resource- and energy-intensive industries, e.g., power and cement, the steel sector has been pushed hard to shift towards a quality-over-quantity and a low-carbon development path. Initial upgrading results were remarkable. Energy consumption per ton of crude steel produced decreased significantly from 605 to 527 kgce (kilograms of coal equivalent) during the 12th FYP (2011–2015), indicating a great improvement in terms of energy efficiency.

1.3 Energy Consumption and CO_2 Emissions

In China, fossil fuels are still the primary source for energy and raw material supply utilized in the industrial sector such as power, cement, steel, petrochemicals, and etc.. Among various industrial activities, the iron/steel sector is one of the most energy- and coal-intensive sectors and emits a large amount of CO_2 into the atmosphere, accounting for about 15% of total industrial CO_2 emissions in China, and about 60% of carbon emissions from global steel industries. When further breaking iron/steel sector down to detailed sectors, the power, steel and cement sectors in total represent over 75% of carbon emissions in China, generating 45%, 15%, and 15% carbon emissions, respectively.

The increasing trends of energy consumption in China for the total economy, the industrial sector, and the steel sector are shown in Fig. 3.

Additionally, Fig. 4 depicts the trends of CO_2 emissions in China for the total economy, the industrial sector and the steel sector.

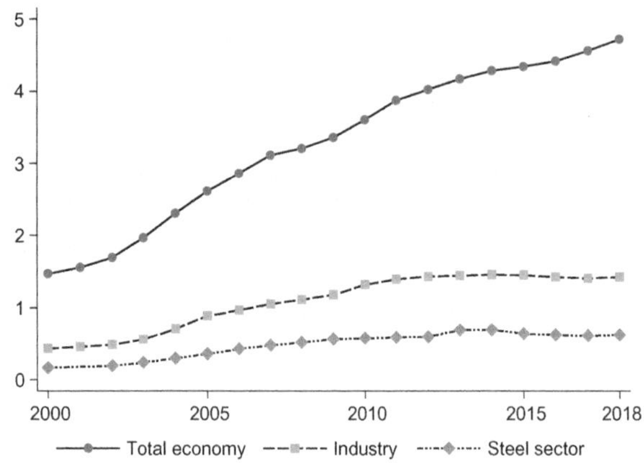

Fig. 3 Energy consumption in China (coal equivalent, billion tons). *Source* IEA (2020)

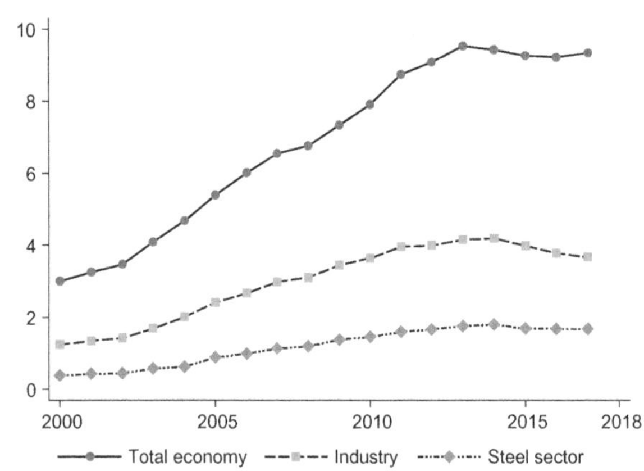

Fig. 4 CO_2 Emissions in China (billion tons). *Source* Shan et al. (2018, 2020)

During China's 13th FYP (2016–2020), "low carbon" and "green" growth has been emphasized as a crucial policy goal, and a series of policy tools were launched to urge industrial firms to take action in saving their energy and cutting CO_2 emissions. For the steel sector in particular, the de-carbonization strategy consists of following three paths: (1) optimization of processes, (2) phasing-out of inefficient capacities, and (3) deploying new technologies at energy conservation and emission reductions.

2 Section 2

2.1 Climate Policy in China: An Evolving Framework

China has made remarkable progress in mitigation and adaptation to climate change in the past decade. The 12th Five Year Plan (FYP) (2011–2015) has identified the low-carbon and green growth as a strategic pathway towards sustainable growth since. The policy framework, however, was not put into place overnight. It has been an evolving process. Indeed, a salient feature of that process was "testing it out before scaling it up" with a mixture of pilots, (special) policies, regulations, and economic incentives. In other words, the policy framework consists of both "sticks" and "carrots" (see Table 1).

Given the complexity of a country's institution of political economy, reform policies have been involving and will continue to involve in various government departments, both hierarchical and horizontal. The top level often provides guidance, either *ex ante* or *ex post*. Action takes place at the bottom level. Successful stories, often not explicitly authorized by the government a priori, could provide basis for formulating new policies *ex post*. Here, the same logic seems to have applied to the formulation of policies for developing a low-carbon and green economy. The need for inter-agency coordination has been particularly salient.

For instance, to launch CCUS pilots would involve multiple authorities, including the National Development and Reform Commission (NDRC), the Ministry of Science and Technology (MOST), the Ministry of Industry and Information Technology (MIIT), the Ministry of Natural Resources (MNR), etc., each with its own distinctive focus. Whereas MOST may focus on the R&D aspects of CCUS, NDRC may be more concerned from the perspective of energy security and economic and social impact from mitigation. And a new ministry, the Ministry of Ecology and Environment (MEE), has been established in 2018 to better respond to challenges posed by climate change.

Table 1 Key elements of climate policy framework in China

Policies with "sticks"			Policies with "carrots"	
Laws and regulations (by central and local governments)			Incentives	
	30 laws:	e.g., Climate Change Law		e.g., pricing; subsidies; loan discounts; fiscal and tax support; access to special funds
	90 regulations:	e.g., Administrative Regulations on ETS		
Administrative orders (by central governments)			Markets	
	e.g., Standards for GHG emissions from industries (by SAC); accounting and reporting methods for GHG emissions in key industries (by NDRC)			e.g., 7 regional ETS pilots; national ETS in power sector; CCERs (by NDRC)

In particular, as mentioned at the beginning of this chapter, there has been a constant and interactive feedback channel between China's international commitments and domestic policy-making in climate change. Under the UN Framework Convention for Climate Change (UNFCCC), in 2009, China had pledged to reduce carbon intensity of GDP by 40–45% by 2020 from the 2005 level, and increase non-fossil fuels as a share of its energy structure to about 15%. Subsequently, the concept of carbon intensity was incorporated into the China's 12th FYP (2011–2015). Similarly, during the Paris Climate Change Conference in 2015, China pledged to peak carbon emissions around 2030, and reduce carbon intensity of GDP by 60–65% by 2030 from the 2005 level. These commitments were integrated into China's 13th FYP (2016–2020). Now that China has pledged to peak CO_2 emissions *prior to* 2030 and become carbon neutral in 2060, one would expect that the coming 14th FYP (2021–2025) will embody a number of policies aimed at peaking CO_2 emissions *prior to* 2030, and begin to work toward achieving carbon neutral in 2060.

In what follows, we review some of the key policies, regulations, and laws, which together have fleshed out what has been and will continue to be an evolving policy framework for formulating a low-carbon and green economy in China, including the topic of CCUS as a negative carbon tool, in addition to low carbon and zero carbon technologies. In the process, among the complex and bewildering network of government agencies, we identify the key government departments tasked with green and sustainable growth in China.

2.2 A Nexus of Legislation, Regulation, and Policy

Currently there are about 30 laws and 90 regulations regarding the development of low-carbon and green economy in China (see Table 1). In 2011, for better synergic work in a complex network of government agencies, the National Leading Group for Climate Change Legislation was set up to coordinate the legislation on issues of climate change. The group is comprised of the Environmental Protection and Resources Conservation Committee, the Law Committee of the National People's Congress (NPC), the Legislative Affairs Office of the State Council—China's cabinet, and as many as 17 central government ministries.

Among the 17 central government ministries, the National Development and Reform Commission (NDRC) is tasked with spearheading the legislative process, including drafting of laws, e.g., the Climate Change Law and the Administrative Regulations on Carbon Emissions Trading.

Climate change-related legislation, regulation, and policy all tended to be very complicated. The making of China's National Plan on Climate Change (2014–2020) gives a good example of the complexity of the planning and implementation process. It involved not only many central government departments but also many local government departments. Indeed, setting ambitious but realistic targets is not an easy job. The effectiveness of policy impact is ultimately determined at the local levels rather than at the central level in Beijing. In other words, for policy implementation, there has to be a feedback channel in the political economy of central-local relations in China.

Indeed, the top-down and bottom-up approach was reflected in the way that China built carbon market, known as emission-trading system (ETS) in Europe. In China, the 12th FYP (2011–15) had provided for "gradually establishing a carbon emissions trading market". In line with the 12th FYP, in 2013, NDRC officially launched seven regional ETS pilots in China, including Beijing, Shanghai, Hubei, Guangdong, Shenzhen, Tianjin, and Chongqing—a topic that will be examined further in Part III of this study. In the same year, NDRC also issued a Circular on Promotion of Experiment and Demonstration in Carbon Capture, Utilization and Storage (CCUS), aimed at promoting CCUS as an emerging technology for direct CO_2 emission reductions. As both CCUS and ETS are important topics of this study, they will be explored in a more systematical way in Part II and III, we will skip further discussion here to avoid redundancy.

It is important to note that in China's evolving policy framework for low-carbon and green growth, "administrative orders", more ad hoc than legislation, play an important role. These policy instruments are particularly salient in the energy conservation and CO_2 emission controls. Examples include performance thresholds, green procurements, safety standards, and industry thresholds. For instance, in 2015 the Standardization Administration of China (SAC) issued 11 national standards for GHG emissions from industries. The Ministry of Transport (MOT), the National Railway Administration (NRA), and the State Forestry Administration (SFA) also issued low-carbon standards. The NDRC provided accounting and reporting methods for GHG emissions in 23 key industries. The National Technical Committee on Carbon Management (NTCC) was responsible for formulating and revising national standards for the carbon management, which includes a system of low-carbon product certification launched in 2015.

2.3 Low-Carbon Policy for Industry

The industrial sector in China is a major energy consumer and CO_2 emitter.[8] Key battles have to be fought there. Indeed, during the 12th FYP (2011–2015), China came up with a series of policies, with both "sticks" and "carrots", all aiming at low-carbon and green growth. In the 12th FYP period, these policies came in the form of 10 plans, 9 programs, 13 opinions, 6 notices, and 8 decisions, measures and guidelines, which fall into six buckets: structural adjustment, standard development, regulatory instruments, economic incentives, market-based instruments, and policy guidance.

The mix of "sticks" and "carrots" of these policy instruments as whole seemed to have worked in reducing energy consumption and CO_2 emissions. For example, while the share of energy-intensive industries declined, that of high-tech industries increased. In 2010, energy-intensive industries accounted for about 30.3% in terms of value added, by 2015, the comparable number was 27.8%, down by 2.5% from the 2010 level. In the process, a large amount of outdated production capacity was removed, including 91 million tons of iron, 95 million tons of steel, 657 million tons of cement, and 2 million tons of electrolytic aluminum. At the

[8] According to IEA classification of 'Industry', in 2015 for example, industry sector alone accounted for 43% of national energy consumption and 52% of national carbon emissions.

same time, energy efficiency was significantly improved. The energy consumption per unit of value added dropped by 28%, representing 690 million tons of standard coal saved. In particular, six major industries—steel, nonferrous metals, petrochemicals, chemicals, building materials, and paper-making—narrowed the gap with their global leaders in terms of energy consumption per unit of products.

2.4 Low-Carbon Policy for the Steel Sector

In 2006, under the United Nations Framework Convention on Climate Change (UNFCCC), China had pledged to the world to reduce energy intensity of GDP by 40–45% in 2020 from the 2005 level. Consistent with that pledge, during the 11th FYP (2006–2010), China had for the first time incorporated the concept of "energy intensity" into the national economic and social development plan. To promote low-carbon pathways, policy-makers started to set binding targets for energy conservation and energy efficiency and took proactive measures to adjust the industrial structure by inhibiting fossil-intensive industries. In the 12th FYP (2011–2015), carbon intensity of GDP was integrated into the national plan for economic and social development, and CO_2 emission controls became an eye-grabbing issue especially for the steel sector.

In retrospect, a total of 36 key policies were issued for the iron/steel sector since 2006, including 12 during the 11th FYP (2006–2010), 17 during the 12th FYP (2011–2015), and 7 during the 13th FYP (2016–2020). These policy instruments can be categorized into five areas: structural adjustment, mandatory tasks, economic incentives, industry standards, and industry guidelines. And the policy making involved a long list of government departments, including the State Council, the NDRC, the Ministry of Industry and Information Technology (MIIT) (for the structural adjustment); the NDRC, the MIIT and the SAC (for the industry standards); the State Council, the NDRC, and the Ministry of Finance (MOF) (for economic incentives); and the State Council, the MIIT, the NDRC and local governments (for mandatory tasks).

While industry guidelines are not mandatory, they have the effect of encouraging industrial firms to shape or rather re-shape strategic planning, carry out R&D, and shift towards green-oriented circular economy via energy saving and technology upgrading, etc.

During the 11th FYP (2006–2010), the policy focus for the steel sector centered on structural adjustments, aiming to enhance energy efficiency, control overcapacity, and upgrade technology. Mergers and acquisitions (M&A) in the iron/steel sector were encouraged to improve industrial concentration. The results

Table 2 Key targets of FYPs

	11th FYP (2006–2010)	12th FYP (2011–2015)	13th FYP (2016–2020)
International pledges	Reduce carbon intensity of GDP by 40–45% in 2020 from the 2005 level; increase non-fossil to 15%	Reduce carbon intensity of GDP by 60–65% in 2030 from the 2005 level	Peak carbon emissions prior to 2030, and achieve carbon neutrality by 2060
Domestic regulatory moves	Mergers and acquisitions in steel sector and energy conservation encouraged	Industrial standards and access to steel sector raised; pilot ETS launched	MOF offered supportive policies to steel sector; 25-member inter-ministerial joint conference set up with approval of SC to address overcapacity; national ETS started
Formation of carbon market	None	7 regional ETS pilots launched in 2013. They cover many industries, including steel sector	National ETS in 2017, covering only power sector

were significant. Under the combined policy implementation, 123 million tons of inefficient iron production capacity and 72 million tons of inefficient steel production capacity were phased out over the five years. They were replaced by large-scale modern facilitates. Among leading steel plants, the share of blast furnaces with over 1000 m^3 capacity increased from 48 to 61%, and that of converters with over 100 tons capacity increased from 45 to 57%. By 2011, energy consumption per ton of iron dropped to 605 kgce (kilograms of coal equivalent), down by 12.8%, as compared with that in 2005. Table 2 summarizes key targets of FYPs (Five Year Plans).

During the 12th FYP (2011–2015), "low carbon" was identified as a principal objective for the steel sector, and the target of 18% CO_2 emission reduction per unit of value added was set. To this end, industrial standards and access to the sector were raised and new targets were set to further phase out inefficient facilities. As a result, a total of 90.89 million and 94.86 million tons of inefficient iron and steel production capacity were slashed off respectively, although overcapacity continued to exist. Crude steel capacity utilization, for instance, was down from 79 to 70%, and the debt rate of large and medium-sized firms exceeded

70%. In the meantime, although the energy consumption per ton of steel declined from 605 to 572 kgce, it was not adequate to offset the net increase of energy consumption with continued production expansion in the sector.

Efforts for a low-carbon economy had intensified in the 13th FYP (2016–2020). Due to an integral part of quality-over-quantity growth strategy, supply-side reforms were highly emphasized on the agenda of the 13th FYP. In line with this, the overall policy on the steel sector tightened up, where the cumulative reduction in its production capacity was up to 100–150 million ton (including 90 million tons of outdated production facilities). To facilitate the low-carbon economy process, in addition to "sticks", "carrots" were also offered. The MOF issued several policy documents, covering fiscal and tax supports, subsidies, employee relocation benefits, land use, etc. In addition, a mechanism of coordination of 25-member inter-ministerial joint conference was set up with the approval of the State Council. The policy goal was to solve the issue of overcapacity in iron, steel, and coal production.

The 12th FYP had provided for "gradually establishing a carbon emissions trading market". Along with the experiences of the seven regional ETS pilots launched in 2013, during the 13th FYP (2016–2020), China has started the process of building national carbon market by the end of 2017. According to the 2015 China-US Joint Presidential Statement on Climate Change, the national ETS in China would cover some major polluting industries, like steel, electricity, chemicals, building materials, paper-making, and non-ferrous metals. Yet, when China first kicked off the national ETS in 2017, the national ETS only covered the power sector, with the expectation that other industrial sectors would eventually be included into the unified national ETS. At the same time, China has adopted a two-track system of ETSs—a national ETS and seven regional ETS pilots. The former covers only the power sector, whereas the latter covers a wider range of industrial sectors. Some have included the steel sector. Such heterogeneity may reflect the regional diversity of China in sectorial and industrial structures.

3 Section 3

3.1 Climate Policy Regarding CCUS: A Global Perspective

In 2018, the Intergovernmental Panel on Climate Change (IPCC) issued the Special Report on Global Warming of 1.5 °C (SR15). The Report warns that the impact of above 1.5 °C of global warming will spell disasters to the ecosystem on Earth, and highlights the urgency for the global community to undertake CO_2

emission reductions with best available technologies (BATs). In this context, the International Energy Agency (IEA) takes carbon capture, utilization and storage (CCUS) as an integral part of the lowest cost route to achieving climate targets (IEA, 2019b; IPCC, 2018), and removing CCUS from the BAT portfolio would lead to a doubling of costs in the future—the biggest rise in costs from excluding any technology.

While many governments have recognized the role of CCUS as an emerging technology for direct reduction of carbon emissions on industrial scale, actions in terms of policy supports are yet to come. Indeed, many are watching and waiting, given the fact that globally CCUS is still in nascent state with many uncertainties and risks, leading to an investment dilemma. A higher perceived risk means a higher rate of return expected, but making the funding process more difficult. The immediate challenge is that while an operational CCUS project may capture millions of tons of CO_2 emissions per year, it requires hundreds of millions of dollars for an initial investment. As a result, funding has been a big barrier to the development of CCUS. That said, in the past few years, six countries—the US, the UK, Canada, Norway, Japan, and China—have stood out as CCUS leaders in the field (IEA, 2019b).

In what follows next, to better understand where China stands in terms of policy and practice in the field of CCUS, let us take a global fieldtrip first, looking at climate policies and CCUS projects in different countries/regions around the world. After the tour, we will come back to China in Section 4 for reviewing its policy and practice with respect to CCUS—a strategical reserve technology under development in its efforts of mitigation and adaptation to climate change down the road.

3.2 A Global Fieldtrip

The United States

As early as in 2006, the US Climate Technology Program had suggested controlling GHG emissions through CCUS. Accordingly, the US Carbon Capture and Storage Early Deployment Act (2009) encouraged the early deployment of CCUS, and the American Clean Energy Leadership Act (2009) further clarified the regulatory framework and financial assistance for CCUS. In its Clean Power Plan, the carbon emission standards pushed forward low-carbon technologies and gave strong impetus to mitigation in the power supply sector. As CCUS technology remains high on the learning curve, incentives focus on providing for early stage initiatives and reducing long-term uncertainty.

Specifically, the Department of Energy (DOE) incentivized CCUS demonstration projects in the industrial sector with US$ 1.52 billion in the deployment funding, derived from the 2009 American Reinvestment and Recovery Act. Section 45Q of the Internal Revenue Code also provided for carbon sequestration tax credits. Section 45Q gave $20 tax credit per ton of CO_2 captured and stored, and $10 per ton of CO_2 captured and used for enhanced oil recovery (EOR). In 2018, the United States Congress further increased tax credit to $35 per ton of CO_2 sequestered through EOR and to $50 per ton of geological sequestration if the project begins construction by 2024, which help to encourage the early deployment of CCUS. The value of up to $50 per ton of CO_2 over the next five years is expected to spur a new round of CCUS facilities (Department of the Treasury, 2018).

In the United States, there are a lot of local actions at the state level as well. For example, after California passed its legislation on zero-carbon energy mandate and carbon neutrality by 2050 target, other states followed suit. Louisiana, Montana, North Dakota, and Texas have also offered tax incentives for the deployment of CCUS (Clean Air Task Force, 2019).

So far, the United States has the largest CCUS facility in the world, with a cumulative capture and injection capacity of more than 150 million tons. Of the 18 operating CCUS facilities worldwide, 10 are located in the US (including a CO_2 injection capture facility in Canada) (Global CCS Institute, 2019). Among these, the largest is the Century Natural Gas plant and CCUS plant in Texas, with an annual capacity of 8.4 million tons, followed by the Shuter Creek CCUS facility, which has an annual capacity of 7 million tons. The industrial CCS facility in Illinois reached the million-ton mark by extracting CO_2 from an ethanol plant run by Archer Daniels Midland and storing CO_2 deep underground near the plant.

The Petra Nova is the only large CCUS power plant in the United States, and one of the only two in the world. The plant uses post-combustion capture technology to extract 90% of CO_2 emissions from the flue gas of the generator set. The facility captures up to 1.4 million tons of CO_2 emissions on a yearly basis.

Canada and South America

Across the Americas, interest in CCUS remains high. Indeed, more than 10% of Canadian government R&D spending has been directed toward CCUS since 2010. Since then, the country has aimed to drive the development of CCUS through both grant support and emission performance standard. At the provincial level, Alberta and Saskatchevan have a royalty relief protocol for CO_2-EOR, and British Columbia has a carbon tax of US$ 30/ton CO_2 from anthropogenic combustion emissions.

Under the Canadian Sustainable Growth and Climate Action Plan, each Canadian province and territory is required to submit to the federal government with a summary of its carbon pricing plan. Costs must start at Canadian $10 (or more) per ton of CO_2 and rise to Canadian $50 per ton by 2022. The federal carbon pricing scheme took effect in January, 2019, and all direct carbon pricing revenue will return to the jurisdiction of origin (Government of Canada, 2008).

In the Americas, through the Canada Strategic Innovation Fund, several milestones were set as the Canadian government. They has invested $25 million in Carbon Engineering (a Canadian Direct Air Capture company), in partnership with some private firms such as BHP, Chevron, and Occidental. Since 2013, the Petrobras Santos Basin Pre-Salt Oil Field CCS has been on-site to separate and capture CO_2 as part of natural gas extraction. Deploying EOR (enhanced oil recovery) technology, the captured CO_2 is injected directly into the natural gas field. The Boundary Dam CCUS project has the capacity to captures about 1Mtpa CO_2 from a coal-fired power plant and use part of it for EOR purposes at the Weyhurn oil field, with a small proportion for dedicated geological storage. The Quest CCUS facility, a joint venture of Shell (60%), Chevon (20%) and Marathon Oil (20%), captures, transports, and stores over 1Mtpa of CO_2.

Elsewhere, coal and gas power plants, as well as stakeholders in the ethanol market, are keen to speed up CCUS development. In Mexico, there are two CCUS pilots—CO_2 Capture Pilot Project and CO_2-EOR and Sequestration Pilot—underway with the support of the CCUS Trust Fund of the World Bank. These CCUS projects were expected to begin operation in 2020.

The European Union

As early as in 2006, the Green Paper: A European Strategy for Sustainable, Competitive and Secure Energy was issued by the European Union (EU). The Green Paper identified CCUS as one of the three policy priorities to address the challenge associated with energy security and climate change. In addition, the European Strategic Energy Technology Plan (SET-Plan) published in 2007 proposed a dedicated policy to accelerate the development and application of low-carbon technologies.

With respect to CCUS technology, the EU CCS Directive specifically dedicates Articles 13 and 16 to the issue of leakage, and the Recommended Practice: Design and Operation of CO_2 Pipelines of Det Norske Veritas makes provisions for monitoring tools and incident response. In addition, the EU's Seventh Framework Program for Research and Technological Development released in 2007 proposed funding projects related to CCUS and sets aside 300 million EU ETS (Emission Trading System) allowances to support 10–12 CCUS projects. Moreover, the EU Comprehensive Climate and Energy Package of 2009 set specific targets for ETS and CCUS, as well as for renewables as share of energy consumption to increase to 20% and to cut GHG emissions by 20% by 2020 from the 1990 level.

In 2018, in its Clean World for All—A Long-term European Strategic Vision for a Stable, Modern, Efficient and Climate-Neutral Economy, the European Commission (EC) unveiled its plan for Europe to achieve carbon neutrality by 2050 (European Commission, 2018). To achieve that goal, both ETS and CCUS have a strategic role to play. In terms of ETS, based on new research on EU-ETS, the EU beefed up its market stability reserves and was ready to accelerate emission reductions. More specifically, from 2021, the EU will reduce total emission allowances at an annual rate of 2.2%, as compared with the current rate of 1.74%. In terms of CCUS, a €10 billion Innovation Fund was set up, with an aim to speed up the development of CCUS projects. CCUS was expected to contribute anywhere from 52 million to 6.06 million tons of CO_2 per year by 2050 under different scenarios.

In recent years, significant changes have taken place in the way the CCUS project is designed in Europe. The new emphasis is on integrated and full chain solution rather than individual transportation from CO_2 source to storage site. Shared infrastructure for CO_2 transport and storage sites can reduce unit costs across the CCUS value chain. As a result, most CCUS projects in Europe are designed as an integral part of industrial clusters with hubs and spokes. These CCUS-dedicated industrial clusters are expected to be established in several major industrial areas of Europe: the ports of Antwerp in Belgium, Amsterdam and Rotterdam in the Netherlands, Humber and Teesside in the UK (Global CCS Institute, 2016).

The United Kingdom

In the UK, the Stern Report in 2006 provided a basis for meeting energy challenge and developing CCUS. In its wake, the White Paper on Energy in 2007 put forward process-wide CCUS demonstration in power plants across the country. Further, the Energy Act of 2008 provided a framework for licensing, enforcement and registration of storage sites. The world's first Climate Change Act of 2008 incorporated CCUS into the emission reduction plan for the power industry.

From the perspective of the UK government, CCUS is taken as an important technology in its strategy to transit to a low-carbon economy, and the government therefore provides tax supports to facilitate this strategic transition. Accordingly, the UK Climate Change Levy stipulates that the electric power enterprises enjoy an 80% tax relief as long as they reach certain targets for energy efficiency and emission reduction through CCUS applications. As a part of its Clean Growth plan, the UK government has also pledged to invest $100 million in innovation in CCUS to reduce costs, as well as to pursue international cooperation opportunities.

In 2019, based on the recommendation by the UK Committee on Climate Change, the UK government set the goal of achieving carbon neutrality by 2050. To facilitate decarbonization, CCUS was identified to play an important role in the process.

In March 2019, the UK government set up the CCUS Advisory Group—composed of experts from the business, finance and legal professions—to identify current bottlenecks facing CCUS development (CCUS Advisory Group, 2019). Apart from business models that propose to diversify funding mechanisms of CCUS investment, the Advisory Group also recommended the reuse or repurpose of the existing infrastructure in the country, aiming to reduce CO_2 transportation and storage costs.

Earlier in 2018, the UK government had announced that it would create its first net-zero carbon cluster in the North-East of England by 2040. The cluster aimed to capture six million tons of carbon emissions on a year basis from a gas-fired power plant and other industrial sources and was expected to start operation by 2030. (European Commission, 2019). Beyond CCUS, Drax, the UK's largest power station, has been conducting tests of the BECCS project and has proposed a local hydrogen and CCS network. A large hydrogen and CCUS center were under plan at St Fergus in Scotland—a good location to add hydrogen directly to the power grid, since 35% of UK's gas comes from onshore. In 2018, the project took step forward when the Oil and Gas Authority—the UK's independent regulator and licensing authority for offshore CO_2 storage—issued the first CO_2 assessment and storage license (Committee on Climate Change, 2018).

Norway

Carbon taxes have been applied in a number of Nordic countries, but only Norway has effectively applied a carbon tax to trigger its CCUS investment. As early as in 1991, Norway introduced a carbon tax on carbon emissions from oil and gas production facilities. This move eventually led to two CO_2 storage projects, Sleipner and Snøhvit, both operated by the Norwegian oil companies Statoil. In 2008, the EU ETS included Norway's offshore oil and gas production emissions, which means that the operators of these facilities must also surrender one EU allowance (EUA) for every ton of CO_2 emitted. But if the CO_2 can be permanently stored, both the tax and EU ETS liabilities can be avoided, which provides powerful incentives to spur CCUS development. Consequently, Norway has been a major player in the global CCUS industry, reducing more than 20 million tons of CO_2 through geological sequestration over the past two decades. Sleipner and Snøhvit are the first companies in the world to store CO_2 emissions offshore and the only large-scale operational CCS facility in Europe.

There is also a major CCS proposal that would involve capturing and transporting CO_2 for offshore storage from multiple industrial facilities in Europe. The proposed facility will be operated by Equinor in partnership with Shell and Total, and it will ship and store CO_2 mainly from northwest Europe (IEA, 2016). FEED research (front-end engineering design) has been doing feasibility studies, and a drilling operation is planned to examine the Norwegian continental shelf for CO_2 sequestration. Based on past success with the large-scale CCS deployment, the Norwegian government intends to use the proposed project as a catalyst for a wider CCS implementation in Europe (Overa, 2019).

Indeed, the Norwegian government has rich experience in the CCUS industry. To illustrate, the government had set up Gassnova, a state-owned enterprise, which has represented the country's investment as a shareholder of Technology Center Mongstad, or TCM. TCM is the world's largest center for testing and developing CCUS technologies. It has the capacity to capture up to 100,000 tons of CO_2 emissions on a yearly basis, and has participated in the first full-scale CCUS project in Norway.

Middle East and Central Asia

There is a growing recognition in the Middle East that CCUS will play an important role in supporting energy demand and shifting to clean energy. Indeed, led by Saudi Arabia and the United Arab Emirates (UAE), each country in the region has a large CCS facility. Whereas Abu Dhabi extracts CO_2 from steel production, Saudi Arabia extracts CO_2 from natural gas production with EOR technology. Abu Dhabi National Oil Company (ADNOC) has recently announced that CO_2 emissions will be reduced by 10% by 2023, and ADNOC also plans to scale up CCS deployment in the next 10 years to capture 5 million tons of CO_2 emissions by 2030 (Adnoc, 2019).

In 2019, Qatar announced that it plans to capture and store 5 million tons of CO_2 emissions from its facilities by 2025 (Zawadzki & Kravtsova, 2019). Qatar Petroleum also announced that it will build the largest CCS-EOR capture and storage facility in the Middle East and North Africa, with a capacity of 2.1 million tons of CO_2 (Aljazeera, 2019). Saudi Arabia and the UAE are members of the Ministerial Meeting on Innovation Tasks and Renewable Energy. Both countries have pledged to double public investment in the renewable energy research and development, and are active participants in the clean energy ministerial CCUS initiative.

Central Asia is one of the world's most energy-intensive economic zone, and it will face enormous challenges to decarbonize the region's energy system and to meet global climate goals. For example, in Kazakhstan, coal, oil, and natural gas have provided 98% of the country's primary energy supply (IEA, 2019a). Given the country's significant coal reserves and dependence on mining, Kazakhstan has a strong interest in CCUS technology. Eurasia Resources Group (ERG), a natural resources company, is currently studying how CCUS can be used to reduce CO_2 emissions from generating electricity from fossil fuels. The preliminary proposal is to retrofit a coal-fired power plant with a 20 mt/year capacity via a CCUS pilot.

Asia Pacific

Over half of the world's total CO_2 emissions come from the Asia–Pacific region. The Asia–Pacific economies also produce over half of the world's most emission-intensive commodities, such as iron, steel, and cement (Global Cement, 2017). Many countries, particularly in South-East Asia, already have and are planning to have more fossil-fuel power plants, and these plants will have a life cycle of decades ahead of them. With the inertia to develop a low-carbon strategy and the lack of an effective legal framework to promote investment in CCUS, the challenge to decarbonize these economies would seem to be enormous. That said, it is encouraging to note that in 2019, the Asia–Pacific region is a home to 12 large CCUS facilities in operation or at various stages of development (GCCSI, 2019).

Japan

In Japan, the Act on Promotion of Global Warming Countermeasures of 1998 stipulated that the control of carbon emissions is the shared responsibility of governments, businesses and the general public. Recognizing the importance of low-carbon technologies, the Japanese government supports research on nuclear power, energy efficiency, renewable energy, and clean fossil energy utilization, including CCUS application. Japan's efforts to develop CCUS has received the policy support and guidance of the Ministry of Economy, Trade, and Industry (METI) and the Ministry of the Environment (MOE).

In 2012, the METI decided to carry out a CCS pilot in Tomakomai City, Hokkaido Prefecture. Earlier, the initial preparation work was commissioned by METI to Japan CCS Co (or JCCS) founded in 2008, which can be regarded as a special purpose vehicle (SPV) explicitly dedicated to the development of CCUS (Japan CCS, 2018). By way of the SPV, the Japanese government funded extensive studies on the various aspects—economic, social, environmental, and legal—of CCUS project, including possible CO_2 storage sites. On the basis of JCCS experience, in 2019, the Japanese government announced its long-term plan to decarbonize its economy under the Paris Agreement. In the plan, CCUS was taken as an important mitigation technology, and the Japanese government decided to work with the private sector as well as other governments to implement a series of initiatives aimed at reducing obstacles to the commercial deployment of CCUS technologies (GCCSI, 2015, 2019).

One example of the Japanese government working to commercialize CCS is the Hydrogen Supply Chain program (HESC). The HESC pilot has involved a number of private businesses, e.g., J-Power, Iwatani, Kawasaki Heavy Industries, Marubeni, Sumitomo, and AGL. The pilot aims to demonstrate hydrogen production from coal with CCS. The construction of the gasifier began in November 2019 in Latropus Valley, Australia, and the project was expected to provide the first batch of hydrogen by 2021. If successful, it was hoped, the demonstration could lead to investment decision around the mid-2020s.

Australia

Australia has a good prospect for CCUS applications, with sound legal and regulatory framework and large geological storage capacities. By making renewable energy and low-carbon technologies available for the same investment opportunities, in 2017, the Amendment to the Clean Energy Finance Corporation (CEFC) further removed barriers to CCUS investment. In Australia, CEFC is a government agency charged with overseeing a US$10 billion clean energy plan (GCCSI, 2019).

In 2019, Australia announced that it would build the country's largest and the world's 19th major CCUS project with EOR technology. When fully operational, the project will have a CO_2 storage capacity of 3.4–4 Mt/year—the largest dedicated geologic sequestration facility in the world. As a result, the Latropus Valley has been gaining momentum around the Hydrogen Supply Chain program (HESC) in collaboration with the Japanese government and private firms. Besides, Carbon Net in Victoria, Australia, has also been conducting feasibility studies to develop a commercial CCUS transport and storage site.

India

India, the world's second-largest energy producer, saw carbon emissions increase by 4.8%, and coal consumption increase by 5% in 2018. However, the IEA has predicted that in a clean energy scenario, India can account for 20% of global industrial CO_2 emissions captured and sequestered by 2060. As such, CCUS has an important role to play in India in the years ahead (IEA, 2019b).

For example, Dalmia Cement, an Indian company, has announced that the company will achieve carbon negative by 2040, and that the company takes CCUS as one of the pathways to achieve this goal. In 2019, the company announced that it will build a large CCUS facility in Tamil Nadu, with an annual capacity to capture 500,000 tons of carbon emissions. To implement the plan, the company has been working with Carbon Clean Solutions, which will provide CCUS technology for the project. The company is also considering multiple carbon capture facilities for source-sink optimalization.

We end our global tour here. Table 3 highlights some of the key elements of the policy or legal framework for CCUS in the major regions/countries that we have just visited before heading back to China.

Table 3 CCUS Regulatory Framework in Major region/countries

	Stage	Legal action	Examplar legislation
EU	Identification	Identify CCUS as a preferred option to tackle climate change	e.g., via Green Paper: A European Strategy for Sustainable, Competitive and Secure Energy (2006)
	Finance/Commercialization	Provide finance to CCUS projects and studies	Climate and Renewable Energy Package (2008)
	Institutionalization	form world's first detailed legislation on CCUS	CCUS Directive 2009/31 (2009)
UK	Identification	Implement CCUS demonstration in power plants	Meeting the Energy Challenge: A White Paper on Energy (2007)
	Finance/Commercialization	Identify CCUS as an important technical choice for electric power industry	Climate Change Act (2008)
	Institutionalization	include provisions on agreeing and using CCUS technologies	Energy Act (2010)
US	Identification	Plan to control GHG emissions through capture and storage	Climate Change Technology Program (CCTP) (2006)
	Finance/Commercialization	update Safe Drinking Water Act by including injection wells purely for carbon sequestration	Bill for Regulation for Underground Storage of Carbon Dioxide (2008)
	Institutionalization	provide ownership of stored gas and responsibility of transfer after CCUS projects shut down	House Bill to Include Carbon Storage Act into Legal System (HB259) (2010)

(continued)

Table 3 (continued)

	Stage	Legal action	Examplar legislation
Australia	Identification	Establish unified CCUS framework within jurisdiction of Australia	Regulatory Guiding Principles for Carbon Dioxide Capture and Geological Storage (2005)
	Finance/Commercialization	N.A	N.A
	Institutionalization	provide legality of offshore storage of CO_2	Offshore Petroleum and Greenhouse Gas Storage Act (2006)

4 Section 4

4.1 Reality Check on CCUS in China

Climate change is increasingly becoming the focus of China's developmental agenda in the past decade, so that the issue has been elevated to "the construction of eco-civilization" described by the Chinese government in many of its official documents.

In 2018, the Chinese central government set up a new and dedicated ministry, the Ministry of Ecology and Environment (MEE), and a big part of the new ministry's mission has to deal with the challenge of climate change. Apparently, even if GDP growth was the top concern previously, the measurement is no longer enough. Now the indicators of carbon intensity must be taken into account in the GDP growth, since the single-minded GDP growth will destroy the ecosystem on Earth. As reflected by the evolving policy framework for low-carbon and green growth punctuated by the country's FYPs, the trend has been clear: as the economy grew as a share of the world's total, so does its mitigation and adaptation efforts. So far, the general contour is marked by three milestones: 2020 (carbon intensity);[9] 2030 (carbon peak)[10]; and 2060 (carbon neutral).

[9] In 2009, China had pledged that it would lower carbon intensity of GDP by 40–45% by 2020 from the 2005 level. By the end of 2019, China had successfully surpassed that target by achieving a 48.1% reduction in carbon intensity of GDP. During the same period, non-fossil fuels as a share of its energy structure rose from 7.4% in 2005 to 15.3% in 2019. Based on this success, China has pledged to reduce carbon intensity of GDP by 65% by 2030 from the 2005 level, and to increase non-fossil fuels as a share of its energy structure to 25% by 2030.

[10] By comparison, the EU peaked CO_2 emissions in 2006 and the US in 2007. As part of its NDCs (nationally determined contributions) under of Paris Agreement, of late China has

Currently China produces close to one third of the world's CO_2 emissions and urgently needs to reduce them. The Chinese government has viewed CCUS as a promising or strategic reserve technology for the direct abatement of CO_2 emissions on large industrial scale,[11] and as such, the government has already put in place a workable regulatory and policy framework to shape its development. As it shown in Table 4, while there is no law dedicated to CCUS in China, CCUS projects are nonetheless regulated by new and relevant policies and existing laws, which, collectively and/or individually, cover the various aspects of CCUS projects. In the life cycle of a CCUS project, these aspects would include project approval, ownership, operation, environmental protection, safety standard, and closure process.

To see how these seemingly separate laws and policies work, let us elaborate on the process by setting up an imaginary CCUS project in China. The proposal for setting up such a project will go through the Construction Law and the Administrative Licensing Law to get project approval. Then ownership of the project will get recognized under the Property Law, the Land Management Law, as well as the Contract Law. While the project is in operation, the CO_2 capture is under the purview of the Production Safety Law; CO_2 transport is governed by the Highway Law, the Law on Protection of Oil and Gas Pipelines, as well as the Road Transport Regulations; and CO_2 storage and utilization fall within the purview of the Mineral Resources Law. Meanwhile, the CCUS project has to comply with a variety of environment and safety regulations, such as the Environmental Impact Assessment Law, the Labor Law, the Production Safety Law, the Tort Liability Law, the Regulations on Hazardous Chemicals, and the Regulations for Safety and Supervision of Pressure Pipelines, etc. In the event that the CCUS project causes environmental damage or degradation, the Environmental Protection Law or the Marine Environmental Protection Law will be invoked.

According to the Global Carbon Capture and Storage Institute (GCCSI), by the end of 2016, there were a total of 38 large-scale integrated projects (LSIPs).[12]

---

pledged to peak CO_2 emissions *prior to* 2030. In addition, carbon intensity of GDP will drop to 65% from the 2005 level; 6 billion m^3 of forest will be added to the 2005 level; non-fossil fuels, as a share of its energy structure, will increase to 25%; and capacity of wind and solar power will increase to 1.2 billion kw by 2030.

[11] Integrated with other technologies, CCUS may neutralize CO_2 emission, or may remove CO_2 from atmosphere, thus contributes to carbon neutrality, or even negative emission.

[12] Large-scale integrated projects (LSIPs) refer to large-scale projects that integrate carbon capture, transport and storage, with an annual capture capacity of 800,000 tons for coal-fired power plants and with an annual capture capacity of 400,000 tons for other emissions-intensive sources.

Table 4 Policy Framework for CCUS development in China

Authority	Dedicated Policy Documents	Year
MOST	Science and Technology Road map of China's CCUS Development	2011
MOST	National Special Plan for CCUS Technology Development in 12th FYP	2013
NDRC	Notice on Promoting CCUS Experiment and Demonstration	2013
MEPC	Notice on Strengthening Environmental Protection of CCUS Pilot Demonstration Projects	2013
	Relevant Policy Documents	
State Council	National Guideline on Medium- and Long-term Program for S&T Development (2006–2020)	2006
MOST	China's National Climate Change Program	2007
MOST	National S & T Actions on Climate Change	2007
MOST	National Plan for S & T Development (12th FYP)	2011
State Council	Working Plan on GHG Control (12th FYP)	2011
NDRC	Coal Industry Development (12th FYP)	2012
State Council	Energy Development Plan (12th FYP)	2013
State Council	Opinions on Accelerating the Development of Energy Saving and Environmental Protection Industry	2013
NDRC	National Key Low-carbon Technology Catalogue (I)	2014
NDRC	National Plan on Climate Change (2014–2020)	2014
State Council	Enhanced Actions on Climate Change: China's Nationally Determined Contributions	2015
NDRC	National Key Low-carbon Technology Catalogue (II)	2015
MEE	Opinion to Promote Super Low Emission in Iron/Steel Sector; Guideline for Super Low Emissions in Iron/Steel Sector	2019
MEE	National Measures for the Administration of Carbon Emissions Trading (for Trial Implementation)	2021

Among them, 15 projects became operational and captured more than 29 million tons of CO_2. While most LSIPs are in the United States and China, those already in operation are mostly in the United States and Canada, and they mostly involve natural gas and coal-fired power plants. Likewise, most CCUS pilots in China are targeted at energy and coal-fired power plants, and none has existed in the steel sector so far. However, unlike the United States where Section 45Q of the

Internal Revenue Code has been a potential tool to encourage CCUS development, in China, government supports thus far have been mainly targeted at R&D of CCUS pilots and mostly by the Ministry of Science and Technology (MOST). The latest Road Map for CCUS in China was released in May 2019. Recognizing the important role of CCUS as part of the country's strategy for low-carbon and green growth, the road map sets out medium- and long-term goals and objectives for achieving low-carbon transformation through affordable, feasible, and reliable CCUS technologies (Asian Development Bank, 2015).

In 2019, China has one large LSIP project in operation, two under construction, and five in early stage of development. The CNPC Jilin Oilfield CO_2-EOR demonstration, with a total injection capacity of 1.12 million tons, was the first LSIP in operation in China and the world's 18th LSIP facility. Since 2007, the Sinopec Zhongyuan Oilfield has been injected with CO_2 from a nearby chemical plant with an annual capacity of 100,000 tons. The Dunhua Methanol Plant, which began operation in 2015, could collect 100,000 tons of methanol on a yearly basis. These were all SOEs (state-owned enterprises). In terms of ownership, the Xinjiang Dunhua Petroleum Technology Co. was about the only private company deploying CO_2-EOR technology in China.

After over a decade of R&D in CCUS technology, China has made remarkable progress in low-concentration CO_2 capture, manufacturing of the carbon capture equipment, CO_2 enhanced oil recovery (EOR), and corrosion prevention. For the carbon capture, China has built several carbon capture plants with capacity of over 100,000 tons each. In 2019, China captured about 2 million tons of CO_2 emissions, representing about 0.02% of the total CO_2 emissions nation-wide, and CO_2 captured as of 2019 added up to 3.7 million tons of CO_2 emissions. As for the carbon transport, China has experimented with all three modes, i.e., pipeline, tanker and sea transport, and completed a total of 50 km of pipeline in length. For carbon storage, China has built an onshore saline aquifer CO_2 geological storage demonstration project with a capacity of 100,000 tons/year, which, once included in saline aquifer storage projects, can increase the annual CO_2 storage to nearly 1.1 million tons and the total CO_2 storage to more than 5.6 million tons.

In all, by 2020 China has established about 30 CCUS projects, with extensive demonstration of the various industrial aspects or processes of CCUS technology. It has set up several CO_2 capture projects with a capacity of 100,000 tons or more each, and built CO_2 transport pipelines with a combined length of 50 km. As discussed earlier, so far most of these CCUS projects are in the energy sector and none in the cement or the steel sector.

4.2 Challenges Ahead

The Paris Agreement sets out a global action plan to set the world on track to avoid dangerous climate change by limiting global warming below 2 °C above pre-industrial levels in the long-term, and to pursue best efforts to limit the increase in warming to 1.5 °C. To achieve this target, the Agreement emphasized the need for global GHG emissions to peak as soon as possible and called for the contracting states to the Agreement to reduce CO_2 emissions with best available (BATs).

Generally speaking, there are three categories of BATs, readily available or still on the horizon, i.e., net-low carbon technology, net-zero carbon technology, and net-negative carbon technology. Along this spectrum of technologies of mitigation and adaptation to the impact of climate change, carbon capture, utilization and storage (CCUS) belongs to the last category, *i.e.*, net-negative carbon technology—a topic that Part II of this study will further explore and discuss in more depth.

A reality check of the existing CCS demonstration projects around the world suggests that while China is not too behind on this technology, more must be done to realize its potential in the years ahead now that the country has pledged to peak CO_2 emissions *prior to* 2030 and to achieve carbon neutrality by 2060. More specifically, while China has made remarkable progress in CCUS demonstrations, they are mostly in the power sector but none in the steel sector so far. The fact is that the iron/steel sector in China contributes more than half of the world's total CO_2 emissions from iron/steel making, and about 15% of China's total CO_2 emissions from the industry. And the policy review by this study on climate mitigation and adaptation in China indicates that the iron/steel sector so far has relied mainly on net-low carbon technologies in its strategic shift toward green growth.

Looking forward, the immediate challenge is to explore and indeed, ascertain the best strategic pathways or road maps for the deployment of CCUS to decarbonize the steel sector—a topic that the next chapter of this study will follow upon.

According to IEA, there will be a 50% chance to control the global temperature rise below two degrees by 2050 when the CO_2 emissions are cut by 60% from the 2013 level and emission reductions total 1000 billion tons. By then, CCUS-derived emission reductions have to reach 6.1 billion tons, which implies an average annual increase of 50% from the current level of 2.9 million tons. Specifically, it means that an estimate of 6000 CCS projects, each with a storage capacity of over one million tCO_2, must be deployed by 2050. At present, however, there are only 3 CCS projects with a storage capacity of over one million tCO_2 globally. Given the current status of CCUS, the challenge will be enormous, so will be the potential, especially in China, and for two reasons. First, currently most CCUS projects are in natural gas and coal-fired power plants, and are not yet well deployed in sectors such as steel, cement, and biotransformation industries. Second, of the total industrial CO_2 emissions in China, the power sector accounts for about 50%, steel 15%, and cement 15%, and as high as 85% of the current energy structure in China is based on fossil-fuels, of which coal takes up 58% and oil 19%.

Acronym	**Full spelling**
BATs	Best available technologies
CCER	China certified emission reduction
CCS	Carbon capture and storage
CCUS	Carbon capture, utilization and storage
CNPC	China National Petroleum Corporation
EOR	Enhance oil recovery
ETS	Emission-trading system
FYP	Five Year Plan
GCCSI	Global Carbon Capture and Storage Institute
GHG	Greenhouse gas
IEA	International Energy Agency
IPCC	Intergovernmental Panel on Climate Change
LSIPs	Large-scale integrated projects
MEE	Ministry of Ecology and Environment
MIIT	Ministry of Industry and Information Technology
MNR	Ministry of Natural Resources
MOF	Ministry of Finance
MOST	Ministry of Science and Technology
MOT	Ministry of Transport
NCSC	National Center for Climate Strategy and International Cooperation
NDRC	National Development and Reform Commission
NPC	National People's Congress
NRA	National Railway Administration
NTCC	National Technical Committee on Carbon Management
SAC	Standardization Administration of China
SACC	Strategy Analysis of Climate in China
SFA	State Forestry Administration
UNFCCC	United Nations Framework Convention on Climate Change

References

Adnoc. (2019). *ADNOC reinforces leadership role and commitment to environmental stewardship as it strengthens its position as one of the lowest greenhouse gas emitters in the oil and gas industry*. https://www.adnoc.ae/en/news-and-media/press-releases/2019/adnoc-reinforces-leadership-role-and-commitment-to-environmental-stewardship

Aljazeera. (2019). *Qatar building large CO_2storage plant*. https://www.aljazeera.com/economy/2019/10/8/qatar-building-large-co2-storage-plant

Asian Development Bank (ADB). (2015). *Roadmap for carbon capture and storage demonstration and deployment in the People's Republic of China*. Manila.

China Steel Statistics. Beijing: China Iron and Steel Association (Various years).

CCUS Advisory Group. (2019). *Investment frameworks for development of CCUS in the UK. CAG final report*. http://www.ccsassociation.org/files/4615/6386/6542/CCUS_Advisory_Final_Report_22_July_2019.pdf.

Clean Air Task Force. (2019). *State and utility climate change targets shift to carbon reductions, technology diversity*. https://www.catf.us/wp-content/uploads/2019/05/State-and-Utility-Climate-Change-Targets.pdf

Committee on Climate Change. (2018). *Reducing UK emissions. 2018 progress report to parliament*. https://www.theccc.org.uk/wp-content/uploads/201806/CCC-2018-Progress-Report-to-Parliament.pdf

Department of the Treasury, USA. (2018). *Carbon oxide sequestration credit*. https://www.irs/pub/irs-access/f8933_accessible.pdf

European Commission. (2019). *The potential for CCS and CCU in Europe*. https://ec.europa.eu/info/sites/info/files/iogp_-_report_-_ccs_ccu.pdf

Global CCS Institute (Global Carbon Capture and Storage Institute). (2009). *Strategic analysis of the global status of CCS: 2009*. Melbourne, Australia.

Global CCS Institute (Global Carbon Capture and Storage Institute). (2015). *The global status of CCS: 2015*. Melbourne, Australia.

Global CCS Institute (Global Carbon Capture and Storage Institute). (2016). *The global status of CCS: 2016*. Melbourne, Australia.

Global CCS Institute (Global Carbon Capture and Storage Institute). (2019). *The global status of CCS: 2019*. Melbourne, Australia.

Global Cement. (2017). *Global cement top 100 report 2017–2018*. http://www.globalcement.com/magazine/articles/1054-global-cement-top-100-report-2017-2018.

Government of Canada. (2008). *The global status of CCS, 2028*. http://www.globalcement.com/resources/global-status-report

International Energy Agency (IEA). (2019a). *World outlook 2019*. Paris: IEA. https://www.iea.org/reports/world-energy-outlook-2019

International Energy Agency (IEA). (2019b). *Transforming industry through CCUS*. http://www.iea.org/reports/tranforming-industry-through-ccus

International Energy Agency (IEA). (2020). *World energy outlook 2020*. https://www.iea.org/reports/world-energy-outlook-2020

IPCC. (2018). In V. Masson-Delmotte, P. Zhai, H.-O. Pörtner, D. Roberts, J. Skea, P. R. Shukla, A. Pirani, W. Moufouma-Okia, C. Péan, R. Pidcock, S. Connors, J. B. R. Matthews, Y. Chen, X. Zhou, M.I. Gomis, E. Lonnoy, T. Maycock, M. Tignor, & T. Waterfield (Eds.),

Global warming of 1.5°C. An IPCC special report on the impacts of global warming of 1.5°C above pre-industrial levels and related global greenhouse gas emission pathways, in the context of strengthening the global response to the threat of climate change, sustainable development, and efforts to eradicate poverty. Geneva, Switzerland: World Meteorological Organization.

Japan CCS. (2018). *From Japan to the world—Aiming for the practical use of CCS Technology in Japan*. http://www.Japanccs.com/en/corporae/message.php

Jiang, K., et al. (2020). China's carbon capture, utilization and storage (CCUS) policy: A critical review. *Renewable and Sustainable Energy Reviews., 119*, 109601. https://doi.org/10.1016/j.rser.2019.109601

Li, J. (2011). *Carbon capture, utilization and storage (CCUS) Activities in China*. Sanya, China. http://www.cagsinfo.net/__data/assets/pdf_file/0009/81576/1-3-carbon-capture-utilisation-storage-activities-China.pdf. Accessed March 5, 2021.

Li, X. (2020). *The road map of China's steel industry: Reduction*. Beijing, China: Metallurgical Industry Press.

Liu, Q. (2016). *CCUS development in China: Outlook, projects and policies*. Japan. http://jp.globalccsinstitute.com/sites/jp.globalccsinstitute.com/files/content/media_release/122850/files/S2_CCUS_development_in_China_Liu_qiang_NCSC.pdf. Accessed June 16.

National Bureau of Statistics of the People's Republic of China. http://www.stats.gov.cn

Overa, S. (2019). *Telling the Norwegian CCS story, Northern lights: A European CO_2 transport and storage project*. https://www.globalccsinstitute.com/news-media/events/telling-the-norwegian-ccs-story-part-3-northern-lights-a-european-co2-transport-and-storage-project/

PBL Netherlands Environmental Assessment Agency. (2021). *Energy and climate change*. The Hague. https://www.pbl.nl/en/topics/energy-and-climate-change

Shan, Y., et al. (2018). China CO_2 emission accounts 1997–2015. *Scientific Data, 5*, 170201. https://doi.org/10.1038/sdata.2017.201

Shan, Y., et al. (2020). China CO_2 emission accounts 2016–2017. *Scientific Data, 7*, 54. https://doi.org/10.1038/s41597-020-0393-y

U.S. Environmental Protection Agency. (2021). *Greenhouse Gas (GHG) emissions*. Washington, D.C. https://www.epa.gov/ghgemissions

Website of Aisixiang. (2021, 5 March). http://www.aisixiang.com/zhuanti/175.html

World Steel Association (, Worldsteel, &). (2019). *World steel in figures 2019*. Belgium.

Zawadzki, S., & Kravtsova, E. (2019). *Qatar opens CO_2 storage plant, targets 5 mln T a year by 2025*. https://www.nasdaq.com/articles/qatar-opens-co2-storage-plant-targets-5-mln-t-a-year-by-2025-2019-10-08

Zhang, X. (2014). *Status of CCUS in China*. Changsha, China. https://iea.blob.core.windows.net/assets/imports/events/76/CarboncaptureandstoragedevelopmentsinChina.pdf. Accessed March 5, 2021.

Drawing Strategic Roadmap of Decarbonization for the Steel Sector

2

Jun Fu, Min Wang, and Qimin Cai

The steel sector is one of the largest industrial sources of CO_2 emissions, contributing around 7% of the global energy sector CO_2 emissions (including process emissions) (IEA, 2020).[1] In the overall scheme of this study, one may take China's steel sector as a window through which to see the magnitude of mitigation and

[1] IEA (2020), Iron and Steel Technology Roadmap, IEA, Paris, https://www.iea.org/reports/iron-and-steel-technology-roadmap.

[2] In terms of CO_2 emissions, the power, the steel, and the cement sectors are the three largest CO_2 emitters in China. As a share of total industrial CO_2 emissions, the power sector accounts for about 50%, the steel sector 15%, and the cement another 50%. Compared with the other two sectors, the steel sector is arguably more representative of the complexity that China faces in its decarbonization efforts, if only because it is more diverse in production processes and product portiforlios than those of the power and cement sectors. This would make the MRV (monitoring, reporting and verification) system—a necessary component of carbon market—more difficult. The point is that if a system is ready to deal with complexity, it is relatively easy to deal with simplicity. For discussion on carbon market or ETS (emissions trading system) in China, see Chapter 5 of this study.

In writing this chapter, the authors benefited from two seminars on climate change given at Peking Uiversity by LI Junfeng, formerly director-general of the National Center for Climate Change Strategy and International Cooperation (NCSC) in China, as well as from several talks with LI Xinchuang, President of China Metallorgical Industry Planning and Research Institute.

J. Fu (✉) · M. Wang
National School of Development, Peking University, Beijing, China
E-Mail: fujun@pku.edu.cn

Q. Cai
National Center for Climate Change Strategy and International Cooperation (NCSC), Beijing, China

J. Fu et al. (eds.), *Climate Mitigation and Adaptation in China*,
https://doi.org/10.1007/978-981-16-4310-1_2

adaptation efforts under climate change.[2] Indeed, the steel sector has been playing an important role in the country's strategy for developing a low-carbon economy. To highlight the point, a few more statistics shall suffice here.

In 2020, the world's total crude steel output was 1.864 billion tons, to which China contributed 56.5% (World Steel Association, 2021). On average, an integrated blast furnace steel plant produces approximated 2 tons of CO_2 per ton of crude steel produced. Given these simple yet powerful numbers, it is critical to explore ways to decarbonize the steel sector in China, especially when Chinese central government announced to peak CO_2 emissions by 2030 and work towards carbon neutrality by 2060. One important technological option for direct reduction of CO_2 emissions is carbon capture, utilization and storage (CCUS), especially after the current paraphernalia of technologies at indirect CO_2 emission reductions via energy savings and energy efficiency has increasingly exhausted their marginal effects. We must plan ahead. For over the next 40 years, the ambition is no longer low carbon now but carbon neural.

In line with the above, this chapter looks at the steel sector as a critical part of China's strategy for developing a green economy. The chapter is structured into four sections. Section 1 provides an overview of the growth trends of China's steel sector, in terms of production, consumption, and CO_2 emissions. Section 2 examines various policy instruments by the Chinese government to decarbonize the steel sector, in light of the country's new pledge to cap CO_2 emissions *prior to* 2030 and achieve carbon neutrality in 2060. In Section 3, we introduce a simulation model in connection with CCUS for the steel sector to plan ahead and shift from *indirect* to *direct* CO_2 emission reductions. Section 4 concludes by presenting a roadmap for the development of CCUS in the steel sector, as an indispensable part of the country's strategy to achieve carbon neutrality in 2060.

1 Overview of China's Steel Sector

1.1 Steel Production

Since China started the reform and opening-up strategy in 1978, the economy witnessed tremendous growth with real GDP on average growing around 9% annually, which together with the large size of the economy has led to the explosive growth of steel production in China. The annual production of crude steel in China on average grows around 11% during 1978–2020, increasing from 31.78

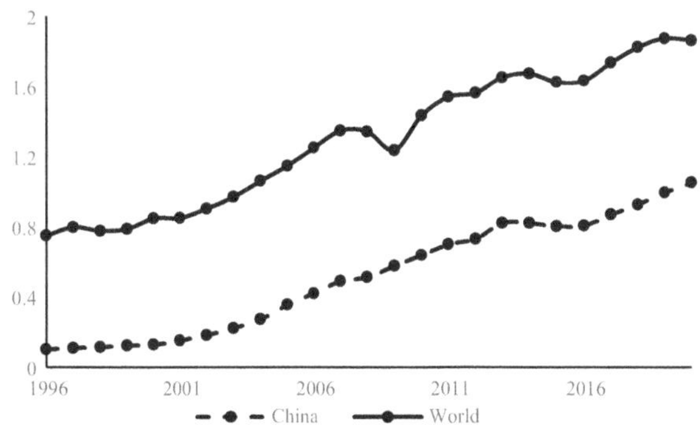

Fig. 1 Total Production of Crude Steel (billion tons), 1996–2020

million tons (around 4.4% of world production) in 1978 to 1.05 billion tons in 2020.[3]

China surpassed Japan as the world largest producer of crude steel in 1996 and has overwhelmingly dominated the worlds' steel industry afterwards by keeping astonishing production growth. The total production of crude steel in 2020 is 10.4 times the total production in 1996. The global share of China's production of crude steel has continuously increased from 15.1% in 1996 to 56.5% in 2020. By contrast, the global share of crude steel production from the second largest producer to the 6th largest producer, EU, India, Japan, Russia and U.S., is 7.4%, 5.3%, 4.5%, 3.9% and 3.9% respectively in 2020.

Figure 1 shows a clear trend of China's production of crude steel in the past two decades. Driven by the new round economic boom especially after China joined the WTO, the growth rate of annual production of crude steel in China reached to a historic high, 21% on average during 2001–2007. After the global financial crisis in 2008, the average growth rate of annual production of crude steel in China fell to 9% during 2008–2013.

Largely due to weak demand, the steel sector was rather dismal during 2014–2016, in which the growth rate of annual production of crude steel was 0.04%, -2.25% and 0.47% respectively in 2014, 2015 and 2016. Chinese steel producers experienced their worst year in 2015, when the capacity utilization rate of the

[3] The data on steel production in different countries used in this paper are from *World Steel Statistical Yearbook* published by World Steel Association, unless otherwise stated.

whole steel sector was only 65% and around half of China's steel mills made loss with many struggling to exit from the sector. The sector has since been under enormous pressures to shift towards a quality-over-quantity and low-carbon development path.

At that time, China government unveiled a series of policies to resolve the steel industry overcapacity problem, such as Opinions on Reducing Overcapacity and Removing Development Obstacles in the Iron/Steel Sector (No.6 [2016] of the State Council). These policies aimed to phase out backward and highly polluting capacity and curtail capacity expansion by setting new higher standards. In total, 100–150 million tons of crude steel production capacity were closed down during the 13th Fifth Year Plan period (2016–2020).

The steel sector recovered quickly after the aggressive policy intervention and is keeping a stable growth rate during 2017–2020. The production of crude steel grows around 6.86% per year in this most recent period. Even when the economy growth was slowed down by the COVID-19 pandemic, the total production of crude steel in China still kept a high growth rate, 5.69%, in 2020. Given the expectation that China may retain average annual real GDP growth rate around 4%–5% in the next three decades (until 2050) and the urbanization rate may increase from 60% in 2020 to 80% in 2050 (United Nations, 2018; Wang, 2020), it is quite likely that China's steel production would keep steady growth in the foreseeable future. In international perspective, in 2020, steel output per capita in China was about 10 tons, 25 tons in the U.S. and 45 tons in Japan.

1.2 Steel Consumption

Figure 2 presents the apparent steel use in China and the whole world between 1996 and 2019. Not surprisingly, as the steel output is mainly consumed by domestic users, Fig. 2 has exactly the same pattern as Fig. 1. In 2019, the global share of apparent steel use of the world's 6 largest steel users, China, EU, Indian, USA, Japan and South Korea, is respectively 50%, 9.13%, 5.76%, 5.74%, 3.69% and 2.94%. Evidently China is not only the largest producer but also the largest consumer, dominating the world steel consumption market.

Moreover, the difference between the global shares of production and consumption of the steel in China also reveals that China is the largest net exporter in the world steel market. It need to note that the total domestic production of crude steel was always lower than the total amount of apparent steel use in China until 2005. That is, China has become a net steel exporter since (including) 2006.

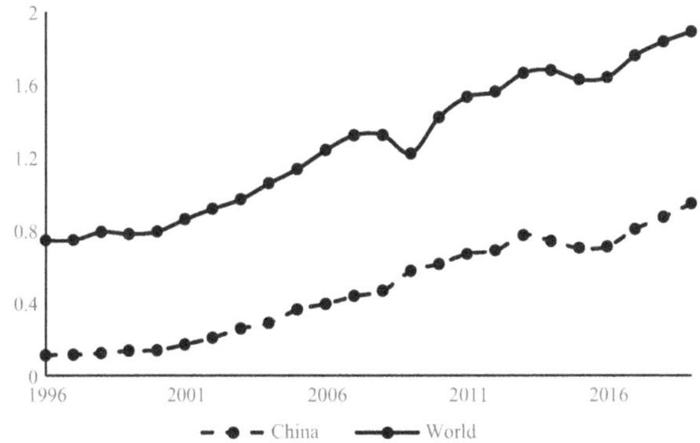

Fig. 2 Steel Use (crude steel equivalent, billion tons), 1996–2019

1.3 Energy Use

Given the explosive growth of steel production and consumption in China, the energy use and carbon emissions from the steel sector are great challenges not only for the sector but also for the whole economy. Fortunately, the energy intensity, measured by the energy consumption per ton of steel, has declined significantly, especially since the 1990s, largely due to policy-induced process restructuring and optimization.

The energy consumption per ton of steel was 2040 kgce in 1980 and almost halved to 1240 kgce in 1999. As shown in Fig. 2, the engery intensity of the steel production fell by 39% during 2000–2019, decreasing from 906 kgce/t in 2000 to 553 kgce/t in 2019. However, given the fast growth of the steel production in the period, the total energy use of the steel sector was increased by 8.6 times, from 116 mtce in 2000 to 996 mtce in 2019.[4] As a consequence, the steel sector has become an increasingly large source of air pollutant emissions and carbon emissions.

Since 2000, under the obligatory energy saving and emission reduction policies, the steel's energy intensity in China kept decreasing significantly until 2006. On average, the energy intensity decreased by 8.6 per year during 2000–2006.

[4] The total energy use is calculated by multiplying the total production of crude steel with the energy intensity.

The decreasing trend of the energy intensity slowed down after 2006. In terms of energy intensity, it only fell at 1.11% per year during 2007–2019 and reached at 553 kgce/ton in 2019. By comparison, the world average energy intensity of crude steel production was 678 kgce/ton in 2019.[5] Thus, the current level of the energy intensity in China is not only much lower than the world average but also close to that in developed countries. That said, both the trend and the level of the energy intensity also indicate that there is limited room for China's steel sector to improve the energy efficiency. Hence, unless there are technological breakthroughs, we expect that the growth of the energy use in China's steel sector would keep pace with the growth of the steel production.

1.4 Carbon Emissions

China has surpassed U.S. as the world largest carbon emitter since 2005. In 2019, China's CO_2 emissions are almost twice of CO_2 emissions in U.S., and China alone was responsible for 29% of global CO_2 emissions (BP, 2020). China's steel sector is a resource- and energy-intensive industry, and the afore-mentioned explosive growth of the production and energy use lead to the rapid growth of CO_2 emissions in the sector. China's steel sector is now emitting around 15% of the country's total CO_2 emissions, right after the largest industry emitter, power sector, which accounts for around 50% of total CO_2 emissions in China. The steel industry is under great pressure to curtain the growth of CO_2 emissions.

It is understood that CO_2 emission from the blast furnaces (BFs) steelmaking process is about around 2–2.2 tonnes for per tonne of steel, while that from the electric arc furnace (EAF) process is far lower at 800 kg per tonne of steel. The BFs steelmaking process (alternatively known as long process) however accounts for around 90% of China's steel capacity. On Dec. 31, 2020, the Ministry of Information and Technology released guidelines for the domestic steel industrial development for 2021–2025. According to the guidelines, EAF steelmaking should be lifted to 15% or 20% by 2025.[6] However, since the products of EAFs do not well fit into the steel demand of manufacturing industries which are the

---

[5] Data source: Sustainable steel—Indicators 2020 and steel applications, World Steel Association.

[6] At the present, EAF steelmaking in China accounts for about 10%; and the rest is long process. By international comparison, EAF accounts for about 47% globally (excluding China). That for many advanced economies such as the US has reached 90%, indicating significant scope for the Chinese steel sector to reduce CO_2 emissions in this area, i.e. from long process to EAF, in the years to come.

key end users in China, there is also limited room for further expansion of the proportion of EAFs.

Recently, Chinese President Xi Jinping announced aggressive actions to tackle climate changes. In particular, China will peak CO_2 emissions before 2030, increase non-fossil fuel around 25% by 2030 and cut its CO_2 emissions per unit of GDP, or carbon intensity, by more than 65% from 2005 levels by 2030. By comparison, the earlier targets that the Chinese government had pledged in 2015 were to peak CO_2 emissions around 2030 (could be after 2030), increase non-fossil fuel around 20% by 2030 and lower carbon intensity by 60–65% by 2030. In line with China's new pledge of climate targets, China has launched national carbon trading market for the power sector in early 2021 and it is expected to cover the steel sector in the near future.[7] And given the fact that the marginal effects of existing technologies on raising energy efficiency and reducing carbon intensity has been on decline, the industry has to explore new ways to decarbonize iron and steel making, such as EAF and hydrogen metallurgy in the short run.[8] But in the long run, CCUS remains another promising option to decarbonize the steel sector.

2 Review of Low Carbon Policy

2.1 Low-Carbon Policy for the Steel Sector

During the 11th FYP (2006–2010), the steel sector in China expanded rapidly, and the policies focused on energy saving and emission reduction. During the 12th FYP (2011–2015), to resolve the overcapacity problem resulted from the great expansion of the steel sector in the 11th FYP, the policies focused on cutting excess capacity, which was regarded as a representative example of "supply-side reforms" at that time. A key component of the 13th FYP (2016–2020) was "eco-civilization", which translated into action plans for industrial firms, means decarbonization and green growth.

---

[7] For more on the development of China's carbon market or ETS (emissions trading system), see Chapters 5 and 6 and this study.

[8] HBIS, China's second largest steel group company in collaboration with TENOVA, an Italian corporation, has been exploring ways to deploy hydrogen technologies in steel-making since 2019. If successful, the new technology can produce steel at a much lower rate of CO_2 emissions, about 125 $kgCO_2$/ton, which is approximately 1/12 of the current rate of CO_2 emissions in long-process steelmaking in China. This is strong indication that, in addition to CCUS, China has been exploring all BATS (best available technologies) to reduce CO_2 emissions.

The policies imposed on the steel sector in the last two decades could be summarized in three categories:

(1) Restructuring the steel industry by resolving the over-capacity. Most policies belonged to this category focused on eliminating existing backward production capacity.[9] These policies generally set a target on existing production capacity that could not meet a certain standard and should be closed down. Some policies might specifically aimed to close down backward processes and backward products.[10] Some policies, such as *Opinions on Reducing Overcapacityand Removing Development Obstacles in Iron/SteelSector* rleased by State Council in 2016, were issued to curtail the expansion of new capacity.

(2) Enhancing the energy efficiency of steel production. The policies in this category generally set the steel sector's environmental standards and energy use standards, such as standards of pollutant emissions, energy intensity, energy saving and energy use in major production processes.[11]

---

[9] Development Policies for Iron/Steel Industry, NDRC, 2005; Notice on Controlling Total Production Capacity, Eliminating Backward Production Capacity and Accelerating Restructuring in Steel Sector, NDRC, 2006; Notice on Accelerating Restructuring Industries with Excess Capacity, State Council, 2006; Notice on Accelerating Industrial Restructuring with Excess Capacity, State Council, 2006; Notice on Comprehensive Work Plan for Energy Saving and Emission Reduction, State Council, 2007; Letter of Responsibility for Closedown and Elimination of Backward Production Capacity, NDRC, et al., 2007; Regarding Curbing Overcapacity, Redundancy, and Guiding Sound Development in Some Industries, forwarded by State Council, 2009; Notice on Further Stepping up the Efforts to Eliminate Backward Production Capacity, State Council, 2010; Notice on Further Efforts to Ensure Energy Conservation and Emission Reduction Targets during the 11th FYP, State Council, 2010; Opinion on Further Efforts to Achieve Energy Saving, Emissions Reduction Emissions, and Acceleration of Restructuring in the Steel Sector, State Council, 2010; Notice on Checking up Iron/Steel Projects, State Council, et al. 2010; Guiding Opinion on Resolving Contradictions of Serious Overcapacity, State Council, 2013; Notice on Effectively Conducting Production Capacity Replacement for Some Industries with Serious Overcapacity, MIIT, 2014; Notice on Formulating Targets and Plans for Elimination of Backward and Excessive Capacity in Key Industries in the 13th FYP, 2014, MIIT, 2014; Notice of Measures for Capacity Replacement in Sectors with Serious Overcapacity, MIIT, 2015.

[10] Industrial Restructuring Guidance–Catalogues. NDRC, 2005; Guidance Catalogue for Elimination of Backward Processes, Equipment, and Products in Some Industries, MIIT, 2010; Industrial Restructuring Guidance Catalogue. NDRC (2011), 2011; Catalogue of Value-Added Tax Preferences for Products and Services in Comprehensive Use of Resources, 2015. State Council, MOF, STA.

[11] The Norm of Energy Consumption per Unit Product of Major Individual Process of Crude Steel Manufacturing Process, 2008, Standardization Administration of China; Access Conditions and Management Measures for Production and Operation of Existing Iron and Steel Enterprises, 2009, MIIT; Norms and Conditions for Production and Operation of Iron and

(3) Changing the energy mix in steel production toward low carbon mix. The policies in this category mainly aimed to promote the development of EAFs technology in steel production, such as proposing the proportion of EAFs that should be used in the steel production and setting an upper limit for energy intensity of the EAF technology.[12]

Many policy initiatives could be traced back to the 12th FYP or even earlier, but they were intensively issued during the 13th FYP right after China government signed Paris Agreement in 2015. Given the above mentioned upgrade of China's 2030 climate change targets, we expect more low-carbon policies will be made and enforced in the steel sector.

2.2 Policy Effects

Under the intensive policy implementation, the steel sector succeeded in cutting excessive capacity dramatically since the 11th FYP (2006–2010). 72.24 million tons of inefficient capacity was slashed during the 11th FYP. Another 94.86 million tons of inefficient steel capacity was phased out during the 12th FYP (2011–2015). During the 13th FYP, 100–150 million tons of capacity, including 90 million tons of backward capacity, were eliminated.

Steel Industry, 2010, MIIT; Guiding Opinions on Mergers and Reorganizations in the Iron and Steel Industry, 2010, MIIT; Design Criterion for Energy Saving of Iron and Steel Industry, 2010, the Ministry of Housing and Urban–Rural Development (MOHURD); Guiding Opinions on Energy Conservation and Emission Reduction in the Iron and Steel Industry, 2010, MIIT; The Notice on Printing and Issuing the Development Strategy and Immediate Action Plan for Circular Economy, 2013, the State Council; The Norm of Energy Consumption per Unit Product of Major Individual Process of Crude Steel Manufacturing Process (GB21256), 2014; Access Conditions for the Coking Industry (Revised in 2014), 2014, MIIT; Norms and Conditions for the Iron and Steel Industry (Revised in 2015), 2015, MIIT; Requirements of the Greenhouse Gas Emission Accounting and Reporting: Iron and Steel Production Enterprises, 2016, NDRC; Code for Design of Energy Saving of Iron and Steel Industry (Draft for Comment), 2017, MOHURD; The Norm of Energy Consumption per Unit Product of Coke (GB21342); The Norm of Energy Consumption per Unit Product of Steel Making Electric Arc Furnace (GB 32,050–2015).

[12] Interim Measures for the Management of Financial Incentive Funds for Energy Saving Technological Transformation, 2007, MOF, NDRC; The 13th Five-Year Plan for the Development of the Scrap Iron and Steel Industry, 2016, China Association of Metalscrap Utilization (CAMU); The Standard of Energy Consumption per Unit Product of Steel Making Electric Arc Furnace, 2016, General Administration of Quality Supervision, Inspection and Quarantine.

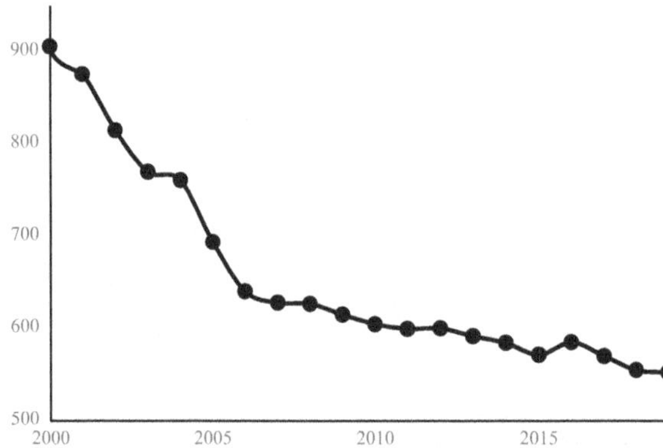

Fig. 3 Energy consumption of Steel in China (Kgce/t), 2000–2019 (*data source* China Steel Industry Yearbook)

In the meantime, more advanced production facilities with better technologies were put into use. As the result, energy efficiency was significantly increased. During the 11th FYP, large- and medium-sized steel plants witnessed a 12.9% drop in energy intensity, falling to 605 kgce in 2010. The energy intensity further decreased to 572 kgce/t in the end of the 12th FYP and 553 kgce/t in 2019. During the 13th FYP, an 18% CO_2 emission reduction per unit of value added was set as a mandatory target in the steel sector.

2.3 Policies Regarding CCUS

Comparing to fossil fuel saving technologies which indirectly reduce the CO_2 emissions, CCUS is the only available end-of-pipe technology capable of directly reducing CO_2 emissions, although it is not commercial yet.[13] The Chinese government takes CCUS as an emerging and strategical technology, and has been actively promoting the development. So far, the government has published more than 26 CCUS-related policy documents at the national level.[14]

[13] We will further elaborate CCUS in a systematical and technical fashion in Chapters 3 and 4.

[14] They included rules, plans, notices and opinions released by the State Council and different ministries and commissions, which can be classified into five categories, namely technology promoting, demonstration support, target setting, environmental management, and others.

During the 12th FYP (2011–2015), the MOST promulgated the National Special Plan for Development of CCUS Technology in 2013. The Plan set out priorities for CCUS pilots and demonstrations. In the same year, NDRC issued a Circular on Promoting Experiment and Demonstration of CCUS. The Circular provided support and guidance for CCUS tests and demonstrations. Also in 2013, MEP issued a Circular on Strengthening Environmental Protection of CCUS Projects. MEP's Circular aimed to enhance environmental protection of CCUS projects, including environmental impact assessment (EIA), monitoring, risk prevention, and damage controls. In 2016, NDRC published the Action Plan for Innovations in Energy Technology Revolution (2016–2030). The Action Plan puts CCUS on its roadmap of key innovations in the energy sector for the 2016–2050 period.

3 Simulation of Deployment of CCUS Technology

3.1 SACC with CCUS

In this section, we present Strategy Analysis of Climate in China (SACC) in connection with CCUS. SACC is a simulation model developed by the National Center for Climate Change Strategy and International Cooperation (NCSC) in China. NCSC researchers[15] collaborated closely with Peking University researchers on this study to figure out the long-term trends of energy use and CO_2 emissions. Key parameters of SACC in connection CCUS in the steel sector are illustrated in Figs. 3, 4 and 5.

The mechanics of SACC-CCUS simulation model works as follows: (1) the model first calculates end energy consumption[16] based on macroeconomic data on GDP, population, industrial structure, and urbanization; (2) the model projects forward with different scenarios of energy mix, given priorities to non-fossil fuels over fossil fuels; (3) the model makes policy recommendations on different paths to achieving carbon neutrality in 2060.[17] For projections, the base year is set at 2015.

[15] NCSC project team was previouoly led by Liu Qiang and is now led by Cai Qimin.

[16] For consistency between model data and China's energy balance sheets, this study incorporates the end energy consumption in agriculture and construction sectors into that in industrial sectors.

[17] Data for the model comes mainly from China Electricity Council, China Energy Statistical Yearbooks, and China Statistical Yearbooks.

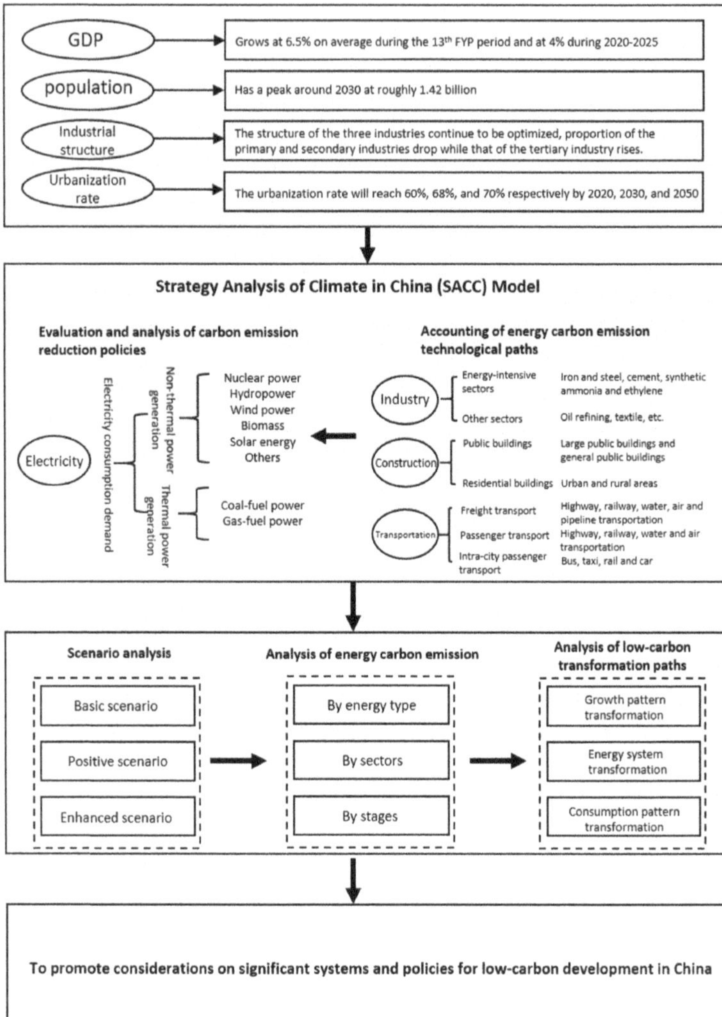

Fig. 4 Framework of the SACC model

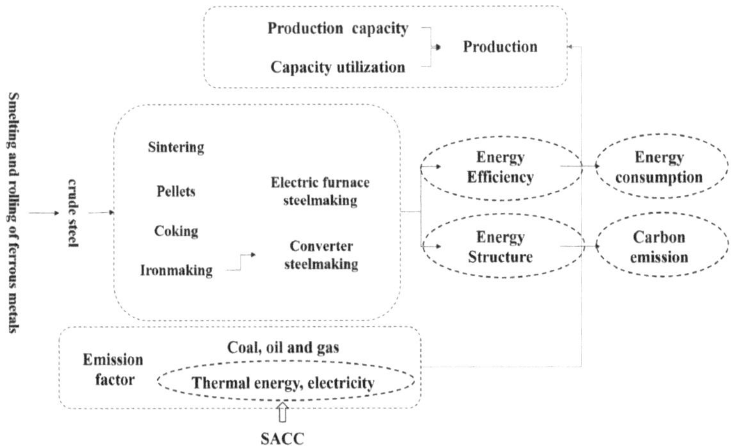

Fig. 5 CO_2 Emission Reduction Model for the Steel Sector. *Source* NCSC

Specific to the steel sector, the SACC-CCUS model takes into account sector-specific variables, which presumably will affect CO_2 emission rates in the steelmaking processes. Key parameters affected by CCUS application to steelmaking processes are circled in red in Figure 10. Here, some of CCUS-specific parameters in the model are calculated as follows:

$$\begin{aligned}&\text{Energy consumption of converter steel making}\\&\quad=\sum\big(\text{energy efficiency of large and medium} - \text{sized processes}_i\\&\qquad\times\text{the proportion of such processes}_i\ \text{efficiency of small processes}_i\\&\qquad\times\text{the proportion of such processes}_i\end{aligned}$$

where *i* represents different processes, namely sintering, pelletizing, coking, ironmaking and converter steelmaking.

$$\begin{aligned}\text{Energy consumption of EAF steelmaking} &= \text{Energy efficiency of large and medium}\\&\quad-\text{sized EAF steelmaking processes}\\&\quad\times\text{the proportion of such processes}\\&\quad+\text{energy efficiency of small EAF steelmaking processes}\\&\quad\times\text{the proportion of such processes,}\end{aligned}$$

$$\text{Carbon emissions} = \sum \Big(\text{energy consumption of converter steelmaking}_j \times \text{emission factor}_j + \text{energy consumption of EAF steelmaking}_j \times \text{emission factor}_j \Big)$$

where j represents different energy types, namely coal, oil, gas, heat and electricity. Emission factors of coal, oil and gas are calculated and estimated based on the The People's Republic of China National Greenhouse Gas Inventory while those of heat and electricity are input from the SACC model.

3.2 Simulation: 3 Scenarios

Based on the SACC-CCUS model of simulation, we set three scenarios of decarbonization for the steel sector, namely a baseline scenario, a low-carbon scenario, and a near net-zero carbon scenario. The basic features of the three scenarios are as follows:

Base-Line Scenario

In the baseline scenario, it is assumed that all energy saving and low-carbon policies introduced thus far will be followed. Specific to the steel sector, all targets set by the 13th FYP will be achieved and low-carbon policies already in place will continue beyond the 13th FYP. But no major technological breakthroughs will be made by 2050, and key low-carbon technologies, including CCUS, are not applied on a large scale.

Low Carbon Scenario

In the low carbon scenario, it is assumed that beyond the 13th FYP, more robust policies will be pursued to reduce carbon emissions than the baseline scenario. Specific to the steel sector, energy conservation and energy efficiency will be further improved. Crucially, the steel sector will go from indirect abatement of CO_2 emission to direct abatement, which means that CCUS will be deployed in the steel sector, as the technology matures over time.

Near Zero Carbon Scenario
In the near zero carbon scenario, it is assumed that all else being equal with the low-carbon scenario, electrification in each of energy end-user sectors will be intensified, and the energy structure will be further optimized. Compared to the baseline scenario, in the steel sector, better production processes such as EAF steelmaking, including an improved auxiliary system of scrap steel recycling, will replace the old long process of steelmaking aggressively, and deployment of CCUS will be taken into full consideration as well.

3.3 Simulation Results

Baseline Scenario
China's carbon intensity per unit of GDP in 2020 and 2030 will decline by roughly 49% and 67% respectively over the 2005 level, in both cases exceeding the targets set by the government. Industrial carbon emissions will peak around 2025 at approximately 6.3 billion tons of CO_2, which will further decrease to 2.1 billion and 2 billion tons of CO_2 in 2035 and 2050 respectively. In the steel sector, carbon emissions will decrease from 1.33 billion tons in 2015 to 1.2 billion, 0.78 billion and 0.46 billion tons of CO_2 in 2025, 2035 and 2050 respectively. By 2050, carbon emissions will be about a third of the 2015 level. Crude steel production will be at a high plateau in 2015–2025 period, and then drop gradually to 0.5 billion tons in 2050.

Low Carbon Scenario
China's carbon intensity per unit of GDP in 2020 and 2030 will decline by about 51% and 71% respectively over the 2005 level, in both cases substantially exceeding the targets set by the government. Industrial carbon emissions will peak around 2025 at approximately 6.5 billion tons of CO_2, and will further decrease to 2 billion and 1.7 billion tons of CO_2 in 2035 and 2050 respectively. In the steel sector, carbon emissions will decrease from 1.33 billion tons in 2015 to 1.1 billion, 0.72 billion and 0.08 billion tons of CO_2 in 2025, 2035 and 2050 respectively. Approximately 0.1 billion and 0.2 billion tons of carbon emissions will be captured by CCUS in 2035 and 2050 respectively. In this scenario, under more stringent policies to control production capacity, crude steel production will gradually decline to the 2010 level by 2030 and drop to 0.45 billion tons in 2050.

Near Zero Carbon Scenario

China's carbon intensity per unit of GDP in 2020 and 2030 will decline by roughly 53% and 76% respectively over the level of 2005, in both cases substantially exceeding the targets set by the government. Industrial carbon emissions will peak around 2025 at approximately 6.3 billion tons of CO_2, which will further decrease to 1.7 billion and 1.3 billion tons of CO_2 in 2035 and 2050 respectively. In the steel sector, carbon emissions will decrease from 1.33 billion tons in 2015 to 1 billion, 0.57 billion and 0.03 billion tons of CO_2 respectively in 2025, 2035 and 2050. Approximately 0.05–0.1 billion and 0.1–0.2 billion tons of carbon emissions will be captured by CCUS in 2035 and 2050 respectively. In this scenario, carbon emissions cut by CCUS, however, may be less than in the second scenario. The reason is that replacement of old steelmaking processes by EAF steelmaking methods will reduce the application of CCUS in the steel sector, although not necessarily for the whole country.

3.4 CO_2 Reduction Potential of CCUS in China

Figure 11 exhibits forecasts by different studies on emission reduction potential of CCUS for China as a whole. The range of emission reductions in the near (2020), medium (2035) and long (2050) term is listed below (Table 11). These studies show that through CCUS, China can reduce 10–200 million tons of CO_2 emissions in 2020 and 100–2240 million tons of CO_2 emissions in 2035. To achieve a higher range of emission reductions, i.e., 1000–3840 million tons of CO_2 emissions in 2050, CCUS will have to be extensively deployed beyond 2035. By 2050, CCUS can be put into commercial use on a large scale, and achieving near-zero carbon emissions means large demand for CCUS technologies (Table 1).

Table 1 Forecasts of CO_2 reduced by CCUS in China ($mtCO_2$)

Year	2020	2035	2050
Range of emission reductions	10–200	100–2240	1000–3840

Sources ADB, Study of China's Carbon Capture and Storage Demonstration and Promotion Roadmap, 2015; IEA, WEO 2015 Special Report on Energy and Climate Change, 2015; Nie Ligong et al., A Study of the Path to Global Commercialization of CCS Technology, 2016; OECD/IEA, Energy Technology Analysis—Prospects for CO_2 Capture and Storage, 2004

Table 2 CO_2 Reduction potential of CCUS in the steel sector

	2030	2035	2040	2045	2050
Emission reduction potential ($mtCO_2$)	10	15	50	75–100	100–200

Source NCSC

3.5 CO_2 Reduction Potential of CCUS in the Steel Sector

In light of the above, and based on our stimulations of low carbon and near-zero carbon shown in Table 11. In the near term, the steel sector will see little emission reduction effects of CCUS. However, the emission reduction potential of CCUS in the steel sector can reach 10 $mtCO_2$ by 2030 and 15 $mtCO_2$ in 2035, accounting for 0.6–15% of the national total, and register 50 $mtCO_2$ in 2040. Up to this period of time, however, CCUS technologies will not have scale effect in the steel sector. However, by 2045, as industrial cluster and infrastructure of transport and storage significantly improve, the emission reduction potential of CCUS in the sector can also increase, reaching 75–100 $mtCO_2$ in 2045, and climbing further to 100–200 $mtCO_2$ in 2050, accounting for 5.2%-10% of the national total (Table 2).

4 Strategic Roadmap for CCUS in the Steel Sector

According to the IEA, to achieve the global energy and climate goals in light of the Paris agreement, direct carbon emissions from the steel sector will have to decrease by more than 50% by 2050 from the 2019 levels, and CO_2 emissions will have to go down to 0.6 tons of CO_2 per ton of crude steel produced, and then move further down towards zero emissions. To make that happen, CCUS has increasingly been recognized as a technology of direct CO_2 emission reduction on industrial scale, and yet, worldwide, the technology, by and large, is still at the early stage of pilots and demonstrations.

In China, the situation is similar. So far there are about 30 CCUS projects under operation or under construction—mostly in energy or coal-fired power plants but none has existed in the steel sector. In September 2020, China pledged to the world to peak CO_2 emissions prior to 2030 and to achieve carbon neutrality in 2060. In view of the new pledge and the significant role of the steel sector in China's low carbon strategy, it becomes increasingly relevant to consider, and indeed promote, CCUS in the steel sector as well. Although very costly now, CCUS is expected to be commercially viable on a large scale by the middle of this century, and when combined with technologies of biomass and EOR (enhanced oil recovery), the technology even has the potential to achieve negative CO_2 emissions. Given the simulation results in the previous section, we develop and indeed recommend a strategic roadmap for CCUS development in China's steel sector. The roadmap has three phases, namely the 14th FYP (2021–2025), 2025–2035, and 2035–2060 (see Fig. 6). We highlight the key points of each of the three phases below:

Phase I (2021–2025)

Phase I overlaps with China's 14th FYP, and it is a preparatory phase on our strategic roadmap. In this preparatory phase, China should get CCUS started in the iron/steel sector on the basis of on-going CCUS projects in other sectors. First of all, by way of by top-level design, China should work out an overall action plan to make sure the iron/steel sector, among many other industrial sectors, peak CO_2 emissions by

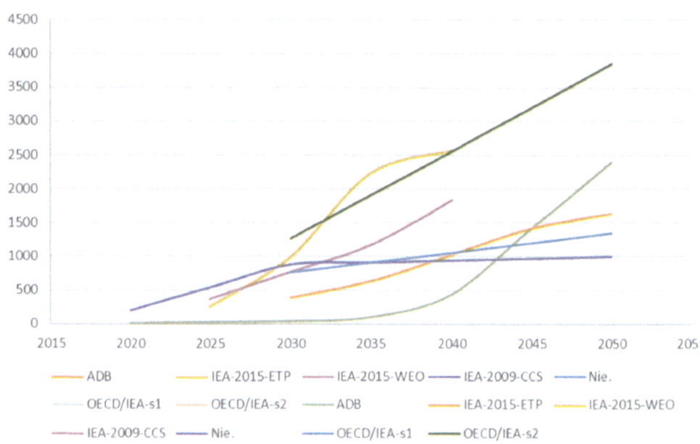

Fig. 6 Forecasts of CO_2 Reduced by CCUS in China ($mtCO_2$). *Sources* ADB, Study of China's Carbon Capture and Storage Demonstration and Promotion Roadmap, 2015; IEA, WEO 2015 Special Report on Energy and Climate Change, 2015; Nie Ligong et al., A Study of the Path to Global Commercialization of CCS Technology, 2016; OECD/IEA, Energy Technology Analysis—Prospects for CO_2 Capture and Storage, 2004

2030.[18] And in line with that action plan, China should increase policy support for large-scale CCUS pilot and demonstration projects. Authorities, both at national and local levels, should explore and develop ways to support CCUS projects.

For example, CO_2 emissions captured and stored by enterprises retrofitted with CCUS should be provided with tax credits and subsidies, including CCER (Chinese Certified Emission Reduction) and be traded in the national ETS. Specific to the steel sector, a few firms should be chosen for CCUS pilot and demonstration projects during the 14th FYP period. In the meantime, public funding to the CCUS-retrofitted firms should go beyond mere R&D money.

Phase II (2025–2035)

Phase II, from 2025 to 2035, will be to consolidate the groundwork for imporved operations of CCUS. The focus during this period should be on three areas: (a) better technologies; (b) enhanced clusters and infrastructure for large-scale full-chain CCUS operations; and (c) improved financing schemes. In terms of technology, China should increase investment in R&D of CCUS, and give stronger support to the integration of CCUS with technologies such as ULCOS and TGR-BF. In terms of infrastructure construction, China should beef up effort to develop CCUS-related industrial clusters, including infrastructure of transport and storage with optimal CO_2 sink-source matching. Specific to the steel sector, during this period, the steel sector should be included in the national ETS so as to open up more channels for financing of, or give more incentives to, the deployment of CCUS for the steel sector. As in Phase I, international cooperation should be encouraged to develop CCUS in China's industrial sectors, including the steel sector, in Phase II and beyond.

Phase III (2035–2060)

Phase III, 2035–2060, is a phase of great expansion of CCUS in China's iorn/steel sector, now that the technogoly is relatively mature, and many CCUS-related clusters and infrastructure are ready for operations. During this phase, a sector-wide technological path for near-zero emissions should be established involving both low-carbon steelmaking technologies and CCUS as technologies of direct reductions of CO_2 emissions. That aside, a new focus of the phase is to enhance international cooperations. This would invovling maintaining international standards, building

[18] A more optimistic prospect is that as an action plan in the making for the 14th FYP, the iron/steel sector has been planning to peak CO_2 emissions by 2025 and reduce CO_2 emissions by 30% from the peak level by 2030.

level-playing field, and exploring ways to include carbon emissions reduced through CCUS in the steel sector in international ETSs, such that CCUS has a better chance to be widely applied in the pursuit of near-zero carbon emissions in the steel sector (Fig. 7).

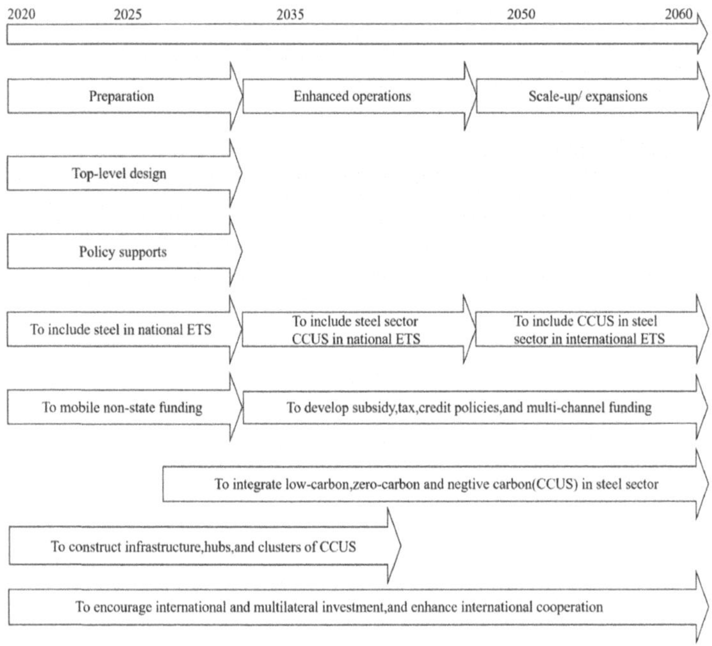

Fig. 7 A phased CCUS Policy Roadmap for Steel Sector

Acronym	Full spelling
BATs	Best available technologies
BFs	Blast Furnaces Steelmaking
CAMU	China Association of Metalscrap Utilization
CCER	China certified emission reduction
CCUS	Barbon capture, utilization and storage
EAF	Electric are Furnace
EIA	Environmental Impact Assessment
ETS	Emission-trading system
FYP	Five Year Plan
MEP	Ministry of Environment Protection
MOF	Ministry of Finance
MOST	Ministry of Science and Technology
NCSC	National Center for Climate Strategy and International Cooperation
NDRC	National Development and Reform Commission
TGR-BF	Top gas recycling blast furnace
ULCOS	Ultra low CO_2 steelmaking
WTO	World Trade Organization

References

Asian Development of Bank. (2015) National Development and Reform Commission of China. Roadmap Research on Demonstration and Promotion of China's Carbon Capture and Storage.

Cai, B., Pang, L., Cao, L., Li, Q., Liu, G., Zhong, P., Zhang, X., Yang, Y., Chen, F., Li, Q., Yang, X., Sun, Y., Li, Y., & Zhang, H. (2019). Two-year implementation assessment (2016–2018) of China's technical guideline on environmental risk assessment for carbon dioxide capture, utilization and storage (on trial). *Environmental Engineering*, *37*(02), 1–7. Global CCS Institute. Global Status of CCS Report: 2019 [EB/OL]. https://www.globalccsinstitute.com/resources/publications-reports-research/global-status-of-ccs-report-2019/. Accessed December 09, 2019.

China Metallurgical Industry Planning and Research Institute. (2019) Low-carbon emissions from the iron & steel industry and prospects [restricted data].

China Metallurgical Industry Planning and Research Institute. (2019). Report on energy conservation and low-carbon development in China's iron & steel industry.

Department of Energy Statistics, National Bureau of Statistics. (2019). *China energy statistical yearbook 2018*. China Statistics Press.

Department of Science and Technology for Social Development, Ministry of Science and Technology, Administrative Center for China's Agenda 21. (2019). *Roadmap for carbon capture, utilization and storage technology development in China*. Beijing: Science Press.

He, X., Huang, Y., & Liao, C. (2014). Review and analysis of CCS policies in domestic overseas. *Advances in New and Renewable Energy, 2*(2), 157–163.

IEA. (2020). *Energy technology perspective 2020 [EB/OL]*. https://www.iea.org/reports/energy-technology-perspectives-2020. Accessed September, 2020

IEA. (2020). *World Energy Outlook 2020 [EB/OL]*. https://www.iea.org/reports/world-energy-outlook-2020. Accessed October, 2020

IEA. (2020). *Iron and steel technology roadmap [EB/OL]*. https://www.iea.org/reports/iron-and-steel-technology-roadmap. Accessed October, 2020

IPCC. *Special report on global warming of 1.5 [EB/OL]*. https://www.ipcc.ch/sr15/

Li, X., Zhang, J., Li, Q., Liu, G., Zhang, X., & Wei, F. (2018). Implementation status and gap analysis of china's carbon dioxide capture, utilization and geological storage technology roadmap (2011). *Science & Technology Review, 36*(04), 85–95.

Li, X. (2019). Roadmap to high-quality development of iron and steel industry in new age. *Iron and Steel, 54*(01), 1–7.

Li, X., & Li, H. Deep decarburization path for high-quality iron and steel development. *China Science Daily*, 003. Accessed November 10, 2020.

Ministry of Science and Technology. (2013). *Notice on issuing the national 12th five-year plan for the development of carbon capture, utilization and storage [EB/OL]*. http://www.gov.cn/zwgk/2013-03/11/content_2351242.htm_. Accessed February 16, 2020

Ministry of *ecology and environment. notice on issuing the technical guideline on environmental risk assessment for carbon dioxide capture, utilization and storage (on trial) [EB/OL]*. http://www.mee.gov.cn/gkml/hbb/bgt/201606/t20160624_356016.htm. Accessed June 21, 2016.

National Bureau of Statistics. (2019). *China statistical yearbook 2018*. China Statistics Press.

National Bureau of Statistics. (2020). *Statistical communique of the People's Republic of China on the 2019 National Economic and Social Development [EB/OL]*. http://www.stats.gov.cn/tjsj/zxfb/202002/t20200228_1728913.html. Accessed March 2020.

National Development and Reform Commission. (2013). *Notice on promoting experiment and demonstration of carbon capture, utilization and storage [EB/OL]*. http://www.gov.cn/zwgk/2013-05/09/content_2398995.htm. Accessed April 27, 2013.

Peng, S. (2013). Implementation of domestic and international carbon capture, utilization and storage (CCUS) projects and development of relevant policies. *Low Carbon World, 1*, 18–21.

United Nations. (2018). *World urbanization prospects* (The 2018 revision, online edition). Department of Economic and Social Affairs, Population Division

World Resources Institute. (2017). *Lessons from international practices for the development of CCSL legal & regulatory framework in China [EB/OL]*. https://www.wri.org.cn/CCS_legal_and_regulatory_framework_CN. Accessed March 2017.

World Steel Association. (2020). *Steel statistical yearbook 2020 [EB/OL]*. https://www.worldsteel.org/zh/

World Steel Association. (2020). *Global crude steel output decreases by 0.9% in 2020 [EB/OL]*. https://www.worldsteel.org/media-centre/press-releases/2021/Global-crude-steel-output-decreases-by-0.9--in-2020.html

Wang, X. (2018). *Theory and practice on energy conservation and emission reduction in China's iron & steel industry under low carbon economy*. Wuhan University Press.

Wang, X. (2020). *Convergence and prospects, china 2049: economic challenges of a rising global power* . Brooking Institution Press.

Yang, H., & Guo, M. (2019). *Opportunities and guidelines for steel enterprises to reduce greenhouse gas emissions*. China: Economic Press.

Zhang, J., Zhang, L. (2020). Preliminary discussion on the development of carbon capture, utilization and storage for carbon neutralization [J/OL]. *Thermal Power Generation*, 1–6. Accessed December 05, 2020.

Part II
Technology

Carbon Capture, Utilization & Storage: A General Overview

3

Dongxiao Zhang, Ruqiang Zou, Xidong Wang, Jin Liu, and Fanyang Mo

With rapid increase in anthropogenic CO_2 emissions, greenhouse effect has been more and more severe and climate change becomes a threat that influences all live beings on earth. In order to mitigate CO_2 emissions, Paris Agreement was signed in 2015 which requires all countries to contribute their efforts in order to limit the global average temperature increase below 2 °C by 2050. The Paris Agreement underlines the urgent need to end the increasing trend of CO_2 emissions and suggests that all technologies, either traditional or novel, should be used in reducing the emissions.

Carbon capture, utilization and storage (CCS) comes as an effective method to solve the problem, especially in fossil fuel-intensive industries. For instance, CCS in steel industry has gain more attention due to the future rising costs of carbon emissions. CCS is projected to account for 33% emission mitigation in industry by 2050, of which more than 80% the carbon dioxide captured is projected to come from steel and cement industries (IEA, 2016).

This chapter gives a general review of carbon capture, utilization and storage, as well as CO_2 transportation. The chapter contains four sections: Section 1 generally introduces the layout of a typical iron/steel plant. Section 2 provides a review of the most appropriate carbon capture technologies in steelmaking. Section 3 covers emerging technologies of CO_2 utilization, including CO_2-EOR (Enhanced Oil Recovery) and BECCS (Bio-Energy CCS). We look at appropriate options for COs transport and storage in Section 4.

D. Zhang (✉) · R. Zou · X. Wang · J. Liu · F. Mo
College of Engineering, Peking University, Beijing, China
E-Mail: dxz@pku.edu.cn

J. Fu et al. (eds.), *Climate Mitigation and Adaptation in China*,
https://doi.org/10.1007/978-981-16-4310-1_3

1 Review of Typical Iron and Steel Plant Layout

Steel production consists of both a long process and a short process. The long process consists of mineral processing, sintering, ironmaking, steelmaking, and the steel rolling process. The short process begins with steel. It refers to compact production process of the combination electric arc furnace, continuous casting and continuous rolling. The main performance is the compact process of raw materials preparation, electric arc furnace melting, refining, continuous casting and rolling Current steel technology gives priority to the long process. The primary technological process in producing steel is shown in Fig. 1. The following sections will describe the long process which includes raw material preparation, ironmaking, steelmaking and steel rolling.

1.1 The Preparation of the Raw Material

1.1.1 Mineral Processing

Mineral processing of low-grade raw ore occurs via the processes of magnetic separation, flotation and re-election, followed by high grade iron concentrate in preparation for the next step of smelting. Main production processes include: crushing iron ore, grinding, mineral processing, drying, and ore concentrate powder.

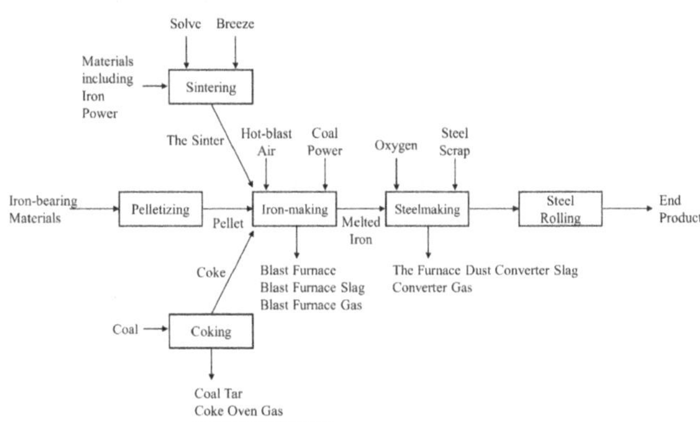

Fig. 1 Main technological processes in producing steel

1.1.2 Sintering and Pelletizing

Sintering production is the process of mixing iron powder, solvent and fuel in specific proportion and then tiling the mixture to a sintering car and ignition. This process causes solution water to evaporate and leads to a series of chemical reactions to occur, ultimately resulting in a partial liquid phase bonding with sufficient strength and granularity. Finally, after crushing, cooling and screening, we produce the finished sinter. Main production processes include burden, first-mixing, second-mixing, sintering, crushing, sieving, cooling and the finished sinter. The production of sintering generates flue gas with high sulfur content, which needs further desulfurization process.

1.1.3 Coking

Coke oven gas is a primary fuel type that is produced by sintering, coking, ironmaking, steelmaking and steel rolling processing and is used for blast furnace smelting. Coking production is the process of mixing and crushing specific coal types, and placing them in a coke oven to produce both hot and crude coke gas through the retorting process. The coking process will produce both coking gas and various other chemical products.

The coke is produced through the carbonization process, in which clean coal is heated in the vacuum-sealed carbonization chamber of the coke oven. The red-hot coke is then sent to a coke quenching tower to be cooled, crushed, sieved and classified. Following this procedure, we are able to get different particle sizes of the coke products.

1.2 Ironmaking

Blast furnace ironmaking is used to make iron ore or iron-bearing materials into the qualified liquid iron by using a reducing agent (such as coke, coal, etc.) at high temperature.

The production process for blast furnace ironmaking is mainly composed of a blast furnace stack and a number of auxiliary systems. The body of blast furnace consists of a furnace throat, stack, bosh, belly, and hearth (moving from top to base of the structure). The auxiliary system consists primarily of feeding, air-supply, coal powder injection control, blast furnace cooling, iron slag and gas purification systems.

The blast furnace ironmaking process begins with the mixing of sinter, pellets, raw ore and other materials in prescribed proportions, sending the mixture to the top of a charging device via a belt or car, and keeping the material surface at a specific height. In this environment, the materials will form ore and coke in alternating layers within the furnace. The coke with then reacts with oxygen, forming carbon monoxide and hydrogen by blowing high-pressure, hot air from the tuyere around hearth. Descending furnace charge and rising coal gas will transfer heat, resulting in reduction, melting, decarburization of the coke to generate the raw iron and slag. Both the molten iron and slag are alternatively expelled from the base of the hearth. The slag and iron are then separated by the skimmer. Molten iron is poured into torpedo ladles or hot metal packages and are sent to steelmaking plants. After the quenching process by water, blast furnace slag can find its use in construction to produce cement. After various processing steps (gravity precipitation, bag dedusting and Top-Pressure Recovery Turbine Plant (TRT) power generation), the coal gas formed at the top furnace can be incorporated into a gas pipe network.

1.3 Steelmaking

The steelmaking process is divided into four processes: molten iron pretreatment, converter steelmaking, refining outside the furnace and continuous casting. The basic tasks of the steelmaking process require the removal of a number of elemental compounds (carbon, phosphorus, sulfur, nitrogen and hydrogen) and the removal the non-metallic inclusions. After this purification process, the material is then heated and molded.

1.3.1 Molten Iron Pretreatment

Molten iron pretreatment is the preparation process for the molten iron of the blast furnace before it is sent to the steel furnace. Without an additional heat source, active substances in the treatment agents will react rapidly with sulfur, phosphorus and silicon. Thus, these elements are removed from the molten iron to form a stable slag phase separate from the molten iron. This process simplifies the steelmaking process and improves the quality of steel.

1.3.2 The Converter Steelmaking

The oxygen converter allows high-speed oxygen to oxidize with the carbon, phosphorus, sulfur, nitrogen and hydrogen in steel to produce slag. The oxidation process also removes non-metallic elements and provides a heat source. The process of steel tapping includes charging and converting the material, deoxidizing steel tapping, slag-splashing for furnace protection and, finally, pouring the slag.

1.3.3 Refining Outside the Furnace Process

Steel refining outside the furnace begins by converting the molten steel produced from primary smelting and pouring the converted steel into ladles. The steel proceeds to undergo deoxidization, desulphurization, decarbonization, and degassing. Non-metallic inclusions are then removed and liquid steel compositions are exposed to high temperatures and undergoes further smelting.

1.3.4 Continuous Casting Process

The continuous casting process is to transform steel from molten state to solid state by use of molds during solid-state phase transformation. The liquid - solid phase change occurs when cool water reacts with the surface of the casting billet, causing the molten steel to undergo the processes of crystallization and casting. In the continuous casting process, the steel will undergo phase transformation and the ingot may experiences bending, correction and some other changes.

1.4 Steel Rolling

The process of steel rolling places rolled pieces of steel into a high-speed roller to generate a desired shape. According to the product types of steel rolling, it can be divided into shaped steel, wire rod, sheet metal and pipe steel. According to production methods, it can be divided into hot rolling and cold rolling.

During the hot rolling process, steel is rolled at temperatures above the recrystallization temperature. This process includes: heating, dephosphorization, roughing rolling, top cutting, precision rolling, laminar cooling, crimp, finishing and plate cutting or heavy rolling.

Cold rolling is steel rolling that occurs at temperatures below the recrystallization temperature. This process includes: dephosphorization with acid pickling, cold rolling, degreasing, annealing, leveling, skin rolling, classifying and packaging.

2 Review of Most Appropriate Carbon Capture and Utilization Technology

2.1 Carbon Capture Technology

The most appropriate technologies used to capture CO_2 can be classified as five types (Carpenter, 2012):

- Adsorption;
- Chemical or physical absorption, or the combination;
- Physical separation;
- Phase separation;
- Chemical bonding.

2.1.1 Adsorption Capture Technology

Adsorption is a mass transfer process where molecules from gas phase are attracted to the surface of a solid material. Adsorption technology in carbon capture involves the passing of a gas stream containing CO_2 through a series of adsorbent beds, where CO_2 are adsorbed. To achieve a desired carbon capture rate, multiple adsorbent cascades are usually used. Once the first adsorbent bed is saturated, the stream is sent to the second bed. The fully loaded bed can be regenerated for recycle use (Wilcox et al., 2014). Therefore, adsorption-based carbon capture is a cyclic process where adsorption and desorption operations are alternately applied. Based on the desorption technique, adsorption technology can be divided into two main categories: the pressure swing adsorption (PSA) (or vacuum pressure swing adsorption, VPSA) and the temperature swing adsorption (TSA) (Olajire, 2010; Wilcox et al., 2014). For PSA, bed regeneration is achieved via reducing the pressure. While for TSA, the adsorbent bed is recycled via increasing the temperature. PSA technology is usually to treat gas stream with high CO_2 concentration, and when the concentration of CO_2 is insignificant, TSA method is favored (Carpenter, 2012; Wilberforce et al., 2019).

Zeolites are crystalline aluminosilicate materials that have a highly ordered porous structure with pore sizes at sub-nanometer scales (Figueroa et al., 2008). Zeolites are widely used as commercial adsorbents for CO_2 capture. Especially, Zeolite-13X is considered as a benchmark for CO_2 adsorption under relatively low temperature (Bui et al., 2018; Hao et al., 2011). Early research has shown that the primary adsorption mechanism for zeolites is physical adsorption, while some studies report that the chemical adsorption mechanism also exists (Spigarelli & Kawatra, 2013). To remove both physically and chemically adsorbed CO_2, the fully loaded zeolites are usually regenerated via a TSA method. The main virtue of zeolites as CO_2 adsorbents are: (1) it has high CO_2 capacity with fast adsorption kinetics; (2) the adsorption performance can be operated under relatively mild conditions (0–100 °C, 0,1–1 bar CO_2) (Spigarelli & Kawatra, 2013). The main drawback of zeolites is that zeolites are very sensitive to impurities such as NO_x, SO_x, and H_2O. These impurities can strongly interact with zeolites, which significantly decrease the uptake of CO_2. Therefore, the flue gas needs to be scrubbed prior to zeolite adsorption process in order to remove any gaseous impurities.

Activated carbon is widely used as a physical adsorption material for CO_2 capture (Lee & Park, 2015). It processes an advantage over other adsorbents for its low-cost and high thermal stability (Spigarelli & Kawatra, 2013). In addition, the pore size of activated can be easily adjusted via changing the material preparation and activation methods (Lee & Park, 2015). The structure of activated carbon can be further functionalized to improve the CO_2 capacity and selectivity of the adsorbent.

Metal oxides such as CaO and MgO are common chemical adsorbents for CO_2 separation. They can isolate CO_2 through the formation of carbonates. MgO, in particular, has the advantages of low-cost, high abundancy, and low toxicity (Lee & Park, 2015). However, compared to zeolite and activated carbon, the CO_2 capacity of MgO is relatively low, especially under dry conditions.

2.1.2 Absorption Capture Technology

Absorption is a chemical or physical process in which molecules of absorbate are dissolved in a bulk absorbent solution. It is a different process from adsorption as absorption is a bulk phase phenomenon whereas adsorption means accumulation of molecules at the surface. Absorption process is a commonly used technology in industry as well as in power plants to isolate CO_2 from flue and fuel gases.

Chemical Absorption

Chemical adsorption is known as the most mature and reliable strategy for CO_2 separation (Leung et al., 2014). It is suitable for large scale CO_2 removal and can be applied even with low flue gas concentration. As CO_2 is an acidic gas, the chemical absorbents used for CO_2 capture are usually amine-based solvents such as solutions of alkanolamines. During absorption process, flue gas containing CO_2 is introduced from the bottom of a packed absorber column, and the absorbent is introduced from the top of the column. When the flue gas encounters the absorbent, they react with each other, forming an CO_2-rich stream. This stream is then sent to a regenerator unit, where CO_2 can be stripped and absorbent solvent can be recycled through heating (100–200 °C). Amine absorption technology (amine scrubbing) with monoethanolamine (MEA) is a commercialized technology and is considered as the most mature CO_2 capture technology in natural gas industry (Olajire, 2010). MEA process benefits from high reaction rate, low cost, and easy amine regeneration. The CO_2 recycle rate can reach to 98% (Olajire, 2010). However, this method suffers from several disadvantages such as (1) limited CO_2 loading capacity (0.5 mol/mol amine) (Mondal et al., 2012); (2) high equipment corrosion rate; (3) amine loss due to SO_2, NO_2, HCl, HF, and oxygen in atmosphere; (4) high energy consumption during solvent regeneration.

Physical Absorption

For physical absorption, CO_2 capture is based on the solubility of CO_2 within the organic solvent, rather than the acid–base reaction. A high-temperature, low-pressure condition is favored to reach to an appropriate CO_2 solubility. A typical absorbent used for physical absorption technology is Selexol. Selexol process has been commercially applied since 1969 to remove both CO_2 and H_2S from natural gas (Olajire, 2010). The absorption process take place at 0–5°C. And Selexol solvent can be recycled via decreasing the pressure or bubbling air through the solution. The advantages of Selexol technology include: (1) low operation cost; (2) low heat rise; (3) gas is dry after selexol treatment; (4) low energy required for solvent regeneration; (5) absorption can be conducted under low pressure. The main weakness of Selexol process include: (1) solvent degradation due to hydrocarbon loss; (2) the process is more efficient under high pressure.

2.1.3 Membranes Capture Technology

Membranes are selective barriers that act as filters to isolate specific gas molecules from the feed matrix (Favre, 2011; Merkel et al., 2010). Membrane technology is a relatively new approach that can be used to separate CO_2 from flue gas, natural gas or hydrogen. It could also serve as another option besides cryogenic method to obtain pure oxygen from nitrogen (Mondal et al., 2012; Olajire, 2010). There are three main possible mechanisms for membrane separation: the Knudsen diffusion mechanism, the molecular sieving mechanism, and the solution diffusion mechanism (Giorno et al., 2015; Inagaki & Kang, 2014). Knudsen mechanism is based on the phenomenon that the gas molecules have more interactions with the pore wall than with each other if the pore diameter is smaller than the mean free path length of the diffusing gas molecules (Giorno et al., 2015). This method can be used to separate gases with different molecular weights. When a mixture of heavy gases and light gases enter the pore, the heavy gases will have more collision with the pore wall and lose more energy than the light gases. As a result, the heavy gases will pass through the membrane slower the light gasses. In this way, gas isolation can be achieved. Molecular sieving method separates gases on the basis of size. Molecular sieves are materials with very small pores. If the size of gas molecule is smaller than the pore size of molecular sieves, the gas can permeate, otherwise, the gas is stopped. The adsorption rate of gases highly depends on the pore size of molecular sieves. In addition, temperature also plays an important role in determining the adsorption rate of gas molecules (Inagaki & Kang, 2014). The solution diffusion mechanism is mainly applied to polymer membranes (Ismail et al., 2005). In solution diffusion model, gases are dissolved in membrane material and diffuse through the membrane, forming a concentration gradient. A separation is achieved owing to the differences in the solubility of gases in the membrane material and the diffusion rate of gas molecules.

Membranes can be made of either organic (polymeric) or inorganic (metallic, ceramic, zeolite, carbon) materials (Olajire, 2010). Generally, polymeric membranes are primarily based on the solution-diffusion mechanism. Some of the polymers that have been studied and applied in membrane technology include polysulfones, polyamides, polyacetylenes, polyanilines, etc. (Lee & Park, 2015; Olajire, 2010). Inorganic membranes mainly rely on the Knudsen diffusion and molecular sieving mechanisms. Inorganic membranes can be divided into two categories: nonporous (dense) inorganic and porous inorganic membranes (Seoane et al., 2015). Nonporous inorganic membranes are generally utilized for selective separation of hydrogen (through palladium alloys) or oxygen (through perovskite materials). Porous inorganic membranes such as carbon and zeolites can be functionalized with a layer of CO_2-philic compound to improve their affinities towards CO_2. Membrane technology benefits from its process simplicity, high flexibility, low energy use, and low requirements for chemical addition (Lee & Park, 2015). However, membrane is not suitable for large-scale filtration. And the durability of membrane material is relatively low. Therefore, a multistage configuration is usually applied in membranes-based system to achieve continuous separation and obtain CO_2 with desired purity (Faruque Hasan et al., 2014).

2.1.4 Cryogenics Capture Technology

Cryogenic distillation method is another commercially available approach for CO_2 separation (Mondal et al., 2012; Olajire, 2010). It is a process that involves cooling and compression of gases, and is commonly used to produce liquid phase CO_2. Cryogenic technology is suitable for capture CO_2 with relatively high concentration. During cryogenic process, CO_2 gas is cooled at a very low temperature and desublimates to solid phase. The solid CO_2 is isolated from the gaseous matrix and the separation is achieved. The frozen CO_2 is then melted with lifting temperature to obtain purified liquid CO_2.

Cryogenic process is a method with high reliability and low investigation cost. It can be applied as an alternative to chemical scrubbing method because it requires less water during operation and generates fewer chemical pollutions. In addition, the cost for cryogenic distillation is relatively low and the device for cryogenic system is more corrosion resistant (Lee & Park, 2015; Wilberforce et al., 2019). As the obtained CO_2 is a cryogenic liquid, they can be easily transported. In fact, in terms of transporting CO_2 by ship, it is most convenient and economically friendly to transport CO_2 in a liquid form (Boot-Handford et al., 2014). Although cryogenic technology offers many advantages, it also suffers from several drawbacks (Olajire, 2010). First, the cooling and heating processes of cryogenic distillation need to consume a great amount of energy, which creates significant energy penalty for carbon capture. Second, impurities such as SO_x, NO_x, and H_2O can have negative impact on the distillation process. These impurities can also cause corrosion and plugging to the equipment. The formation of ice from water molecules can block pipes and result in severe safety problems. Therefore, it is important to remove trace moisture from feed gas prior to the cryogenic approach, which increases the cost.

2.1.5 Gas Hydrate-Based Capture Technology

Hydrate-based CO_2 capture (HBCC) technology is a relatively novel CO_2 capture process. Due to numerous advantages over conventional capture technologies, HBCC has drawn lots of attention from academia and industry. Advantages include mild operational temperature, easy recycling of aqueous solution, allowance of large-scale operation and most importantly, economic advantage (Dashti et al., 2015). The capture mechanism is rather simple. When CO_2 contacts with water under certain condition (below equilibrium temperature and above equilibrium pressure), it forms gas hydrate in which CO_2 concentration is much higher than the gas phase. Subsequently, gas hydrate is separated from the gas and CO_2 is collected from the hydrates in a dissociation reactor, either by heating or depressurization. The water remained is recycled and used in the gas hydrate formation process again. In practice, the whole process is repeated several times to increase the purity of CO_2. Also, chemical additives are needed to reduce the equilibrium pressure (thermodynamic promoters) and increase the hydrate formation rate (kinetic promoters). The chemical additives receive lots of research interests. Propane, cyclopentane, tetrahydrofuran and tetra-n-butyl ammonium bromide are most investigated thermodynamic promoters while dodecyl-trimethyl-ammonium chloride, sodium dodecyl sulphate, and Tween-80 are commonly used kinetic promoters.

2.1.6 Mineral Carbonation Capture Technology

Slags are by-products produced after desired metal has been separated from its own ore. In iron and steel making process, slags are composed of silicates, free lime and other minerals and thus are capable of reacting with CO_2. For example, calcium oxide in the slags could react with CO_2 to form calcite. Such reaction is called mineral carbonation by which CO_2 is captured and permanently stored. Generally, two types of carbonation processes are developing (Baciocchi et al., 2010). The direct carbonation involves CO_2 reacting with slags directly, either in the aqueous phase or at the gas–solid interface. However, in the indirect carbonation process, the metal in slags is firstly extracted with hydrochloric acid (Kunzler et al., 2011), acetic acid (Eloneva et al., 2008; Teir et al., 2007), nitric acid (Doucet, 2010), hydroxides or ammonium salts (Fogelholm et al., 2009). Then CO_2 reacts with extracted metal to form precipitate.

Capturing CO_2 with mineral carbonation has its limitations. First of all, the process has large energy consumption and the solvent regeneration is expensive (Doucet, 2010). Additionally, the whole process itself generates large quantities of CO_2 which makes the capture less efficient. However, the process has numerous advantages. Economically, produced products of the carbonation are profitable. For instance, the carbonates are highly demanded in agricultural and construction applications. Carbonates with high quality would enable its sale to the paper industry (Fogelholm et al., 2009). On the other hand, the process has environmental benefits. Carbonated slag blocks can be put into the ocean to promote the algae growth which will transform CO_2 into O_2 through photosynthesis (Baoshan Iron & Steel, 2008). The blocks can also be used in endangered reef areas to buffer the ocean acidification.

2.1.7 Emerging Carbon Capture Technologies

In recent years, a great number of emerging carbon capture technologies have been reported as promising solutions to reduce CO_2 emission. Some of the innovative low-carbon technologies have already been applied in demonstration scale or are in early commercialization stage (ADB, 2020). For example, solid sorbent technology, as a new CO_2 capture process that can isolate over 90% of CO_2 from the post-combustion flue gas stream, has been commercialized through the ViennaGreeCO$_2$ project, and will be used in pilot plants in Netherlands. A following TuilpGreenCO$_2$ demonstration project is also in progress to upscale this technology to full commercial stage. A CO_2 capture system proposed by Francois Marechal team aims to separate CO_2 from internal combustion engine. This proposed system is still in research stage, but it shows great potential to be applied in Asia and other regions. Another CO_2 capture technology that is highly relevant for implementation in Asia is called the Compact Carbon Capture system. The size of this system can be up to 75% smaller than conversional amine capture systems, which means that the Compact Carbon Capture system can be built in many of the existing power plants where CCS model was not designed in advance. Recently, Carbon Infinity, a company that was founded by research experts at Shanghai Jiao Tong University, developed a direct air capture (DAC) system that enables the removal of CO_2 from gas mixture and convert CO_2 to value-added carbon-based industrial products. DAC system has attracted the attention of the Abu Dhabi government, and there has been increasing interests across Asia in bringing this system to market.

2.1.8 Carbon Capture Recommendation

Carbon capture face some limitations. As can be seen, many capture technologies suffer from high energy consumption and, where used, steam consumption. A potential solution is to utilize waste heat from other on-site processes to regenerate the capture solvents. Another limitation is economical. By far, CO_2 capture is expensive which will hinder its widespread commercial deployment. There are two solutions towards this problem. First of all, promoting large CCS projects could be helpful in cutting costs since large-scale CO_2 capture costs less. Furthermore, developing new technologies is another way forward. For example, HIsarna process, which is designed to generate a nitrogen-free and CO_2-rich off gas which will make CO_2 capture easier and cheaper. If these two limitations are overcome, then CCS projects could be more widely deployed.

3 Carbon Utilization Technology

Millions of tonnes of CO_2 are used in industry each year, and the idea that combining CO_2 utilization with CCS is quite appealing. Instead of disposing CO_2 as a waste, delivering it to willing buyers is a better way on an economic analysis. However, CO_2 utilization could only act as a driving force to develop CCS but cannot replace geological storage. The reason is that the uses are rather limited when considered to be applied to CCS. For instance, the global beverage industry consumes nearly 8 million tonnes CO_2 per year, which is only 0.5% of the CO_2 need to be captured and stored by 2030. Therefore, the CO_2 demand of different approaches should be considered in order to match the capture capacity. The emission reductions and financial contribution should also be analyzed.

3.1 CO_2 Used for Enhanced Oil Recovery

Enhanced oil recovery is the major source of CO_2 utilization, with 70 Mt CO_2 used annually, which is also called tertiary oil recovery. CO_2-EOR technology is injecting CO_2 into an underground geologic zone that contains hydrocarbons to produce the oil. The CO_2 is produced together with oil and then recycled back to recover more oil.

The United States began to conduct relevant projects decades ago and has sequestrated millions of tonnes of CO_2 underground. However, only around one-third of quantities of CO_2 are from anthropogenic sources. Additionally, at present the main concern of CO_2-EOR is to enhance oil recovery. Therefore, EOR technology should be improved when used either primarily or secondarily for CCS. For example, incorporating monitoring and validation systems into EOR could help ensure the long-term retention of CO_2.

Sequestrating CO_2 in reservoirs can not only enhance oil and gas recovery, but could have the added benefit of reducing CO_2 emission. However, when we use numerical models to simulate the process of CO_2-EOR, a problem arises when CO_2 sequestration and EOR should be co-optimized at the same time. Jahangiri and Zhang (2012) proposed the ensemble-based optimization (EnOpt) method to optimize CO_2 storage and oil recovery process simultaneously.

This study takes both CO_2 storage and oil production into account and uses net present value of the CCS project as the optimization objective function. The proposed methodology presented in this work can be achieved with most reservoir simulators and thus designing and co-optimizing the coupled CO_2 sequestration and EOR is possible.

3.2 CO_2 Used in Iron Making Process

Hong (2012) from Northeastern University China applied CO_2 as a desiliconizing agent for hot metal. The experiments were carried out in a laboratory induction furnace, with 1 kg of material prepared using a pig iron with iron silicide with silicon content around 0.745 wt%. CO_2 is injected into hot metal through a quartz tube for desiliconizing under different temperatures ranging from 1300 to 1500 °C, and different gas flow rates ranging between 0.2 and 0.6 L min^{-1}. The results showed that the mass transfer of gas is the rate-limiting step of desiliconizing reaction with CO_2. Under constant temperature, higher gas flow rate will give better desilication. On the contrary, at certain gas flow rate, higher temperature leads to lesser desilication. This is in conformity with thermodynamic analysis since desilication is an exothermic reaction and lower temperature would favour desilication.

Guo et al. (2010) patented the use CO_2 as the transmission medium for pulverized coal injection into a blast furnace instead of N_2 or compressed air. Pulverized coal was first transferred by CO_2 from bins to pipes then injected into BF with CO_2 as the carrier. This was expected to enhance the proportion of CO in the BF gas, hence increase the quality and reducing the coke rate. This is in part due to CO_2 concentrations around tuyeres being very high, and the reaction $C(s) + O_2(g) = CO_2(g)$ will be restrained while the reaction $CO_2(g) + C(s) = 2CO(g)$ will be promoted. One should notice that it is essential to control the CO_2 proportion in the mixed gas, because higher CO_2 will increase the potential energy of the coke reaction. This will result in large amounts of cracks and pores within the coke, decreasing the coke strength. In this case, it will deteriorate the permeability of stock column during the descending process.

Almost at the same time in 2010, Fu et al. (2010) invented a method of injecting carbon dioxide into BF with the aim to reduce the coke ratio. CO_2 is produced from the exhaust gas of the hot blast stove or other exhaust gases which contain CO_2. Replacing the O_2-rich gas stream with CO_2 will result in drastic decreases in temperature, hence the temperature of the injecting gas needs to be increased when introducing CO_2 into the BF. There is no industrial application being reported for the above two patents, maybe because it is difficult to operate.

3.3 CO_2 Used in Steel Making Process

3.3.1 CO_2 Used in Converter Process

Decarburization

The use of CO_2 for decarburization has a history of around 30 years in China. Back in 1985, Jin (1985), an industrial researcher from Shanghai's No. 1 steel plant, reported the positive effects of N_2–CO_2 bottom blowing in a 15 t converter. It was reported that this application can enhance the decarburization efficiency and decrease the total iron oxide concentrations (Σ^{FeO}) in the final slag. However, the corrosion of the refractory around tuyere was found to be severe.

Wang (1986, 1989) carried out a series studies during 1986–1989 on the bottom blowing of CO_2 bearing gases. The high temperature experiments were carried out and the result showed that with CO_2 bottom blowing, the decarburization rate is larger than other gases in the middle period of decarburization. Similar work was carried out elsewhere (Hara et al., 1986; Paules, 1987), aiming to investigate the effect of CO_2 during bottom blowing. The results of these studies showed that CO_2 is an effective alternative for N_2/Ar bottom blowing, since it will not increase the nitrogen content in liquid steel. This would also be a superior alternative to bottom blowing O_2/C_xH_y in converter as the dissolved hydrogen content would not increase during the operation. The mechanism of bottom blown CO_2 in the bath of combined blown converter was investigated in later stage (Guo & Chen, 1993) by Chinese researchers and the industrial test of CO_2 bottom blowing was carried out in a 180 t converter in Anshan Steel Plant (Li & Han, 1996). However, this technology could not be applied in the industry initially due to the short lifetime of the purge plug.

During the experiment, it is found that the corrosion of the tuyere is severe just as Shanghai No. 1 steel's practice (Jin, 1985). It was suggested that, during the progress of decarburization, the carbon content in the melt would be very low and, consequently, the carbon in refractory around tuyere will dissolve into the melt.

Anshan Steel Plant carried out further research to solve the refractory problem and increase the lifetime of tuyere to 1500–2000 times of use. CO_2 bottom blowing in converter had been used widely in this plant until the paper of (Li & Han, 1996) was published. Sodium carbonate (Na_2CO_3) solution absorption method was adopted by Anshan Steel Plant for CO_2 production but the cost of CO_2 preparation has not been reported.

Utilization of CO_2 in metallurgical processes in China has been studied by Jin et al. (2007) and Yi et al. (2009). They started the work around 2004–2005 and named the new process the COMI (CO_2 and O_2 Mixed Injection) process. In 2007, experimental study of steelmaking with CO_2–O_2 mixture was carried out by a 10 kg induction furnace, with the initial carbon content of around 2.5wt%. The result showed that it is feasible to decarburize the melt with CO_2–O_2, and they pointed out that the maximum amount of CO_2 introduction in converter is 13% (Jin et al., 2007).

The process was implemented in 2009 in industrial experiments (Yi et al., 2009) with CO_2 addition ratio of around 7%, and the results were very positive. The consumptions of lime, iron ore as well as oxygen with COMI process were all lower than that with conventional process. Additionally, the Total iron (T_{Fe}) content in slag was decreased by 1.79 wt% compared to the conventional method.

Dust Reduction

One of the important achievements in Chinese research with regards to the benefit of using CO_2 in the converter process is that it can reduce the generated amount of converter dust. The characteristics and mechanism of dust generation was also investigated by Bi et al. (2010). Both laboratory and industrial experiments have been carried out in China (Ning et al., 2009; Yin et al., 2009; Zhu et al., 2010) and the results showed that, with an increase in the CO_2 ratio in the mixture gas, the dust generation is decreased.

Evaporation was considered as the main reason for fine (micro size) dust formation, which constitutes more than 60% of converter dust. The main content of the converter dust is FeO_x (around 50% of total dust). With the oxygen injecting into hot metal, a fire zone forms with temperature higher than 2500 °C (maximum 3000 °C). As the boiling point of iron is 2750 °C, iron will be oxidized and evaporate to form dust at lower temperatures. If one can decrease the temperature of the oxygen jet fire zone, dust generation is expected to decrease. Based on the endothermic nature of the Boudouard reaction, namely the reaction of CO_2 and C, if CO_2 is mixed with O_2 to form the oxidizer gas mixture, the temperature of the fire zone will be lowered so as to decrease the oxidation evaporation of Fe.

The result obtained by Yin et al. (2009) has shown that, compared to the conventional steelmaking process, when CO_2 volume ratio is 5%, the dust amount is decreased by 25%, while at the same time, the T_{Fe} content in the dust is reduced by about 15.3 wt%.

Dephosphorization

The dephosphorization of BOF with the introduction of CO_2 has been widely studied (Lv et al., ; Zhu et al., 2012). The dephosphorization of hot metal with high-Si and high-P in the converter is a serious technical problem during the steelmaking process. Normally, in order to enhance the dephosphorization efficiency, solid coolant is added during dephosphorization with the aim to control the temperature increase rate. However, solid coolant will lead to a partial-cooling of the bath. Consequently, the cooling effect would be non-uniform. Additionally, the impurity in solid coolant will affect the quality of steel production. In this case, CO_2 could be a better choice as a temperature controlling agent. A 30 t converter was adopted for carrying out the CO_2 cooling experiments (Yi et al., 2009).

The industrial trials showed that, with the introduction of CO_2 (namely COMI process), the temperature of the dephosphorization period has been controlled better compared to the conventional process and CO_2 bottom blowing is found to be beneficial for dephosphorization. The temperature of semisteel is around 1330–1350 °C, and the phosphorous content in semisteel is lower than 0.050wt%. The average dephosphorization efficiency for semisteel using conventional processes is 53.10%, while it is 66.49% with COMI process. The P content for the first tapping is also decreased by 23%, from 0.030 to 0.023 wt%.

To summarize the effects of CO_2 used in converter process, the introduction of CO_2 in the proper amount will be favourable to decarburization, dust reduction, reduction of materials consumption and dephosphorization. However, the amount or ratio of CO_2 mixing must be calculated carefully, as too much CO_2 can lead to skull of slag because of the endothermic reaction between CO_2 and C.

3.3.2 CO_2 Used in AOD Process

The application of CO_2 in stainless steel making is extremely interesting. Currently, stainless steel is mostly produced through the argon oxygen decarburization (AOD) process. For ordinary stainless steel production, the Ar consumption is 11-12m^3/t, O^2 consumption is 14-24m^3/t, while, Ar consumption is 18–23 m^3/t in the case of smelting extra-low carbon stainless steel (Chen et al., 2008). It has economical profit if Ar could be replaced by cheaper gases, like N_2, steam, CO_2, etc. However, N_2 will bring N and steam and will take H into the melt, both of which are harmful for certain steel grades. The use of CO_2 could be ideal considering this aspect. The present authors have investigated the feasibility of CO_2 application in the stainless steel making process, both in Sweden and in China.

Wang et al. (2009, 2010) have carried out induction furnace experiments at the Royal Institute of Technology Sweden (KTH) during 2007–2010 and the results show that CO_2 is effective for decarburization without the loss of chromium. After that, Zhu et al. (Li et al., 2015a) have carried out basic study of thermodynamics of CO_2 applied for stainless steel smelting, and pointed out the maximum CO_2 used in AOD process is 9.13%. They stated that for carbon content over 0.5 wt%, it is possible to use $CO_2 + O_2$ mixture, while, at carbon level lower than 0.5wt%, it is better to use Ar to decrease partial pressure of CO (P_{CO}). They also implemented experiments in China (Bi et al., 2012) and found when the ratio of CO_2 increased from 0 to 50%, the carbon content will decrease by 12% while Cr content will increase by 19.93%. These authors have suggested the use of 5–11 m^3 CO_2 for one ton stainless steel production. The price of Ar is around 5–8 times that of CO_2. Consequently, the production cost will be decreased by 3–7 US$/t.

Anderson et al. (1990) have shown, from their corresponding trials at a steel plant in U.S. with production capacity of 1.0 million tonnes per year, that adopting a similar practice as proposed in Bi et al. (2012), 20 Million to 45 Million RMB could be saved per year. In 2011, Anshan Iron and Steel Co., Ltd. (Tang et al., 2011) invented an AOD process with injection of CO_2 to produce stainless steel and they found it is suitable for smelting steel with carbon content from 0.001 to 0.3 wt% using the new process. They also found with CO_2 injection, oxygen lance can be cooled and its life improved by 20%.

3.3.3 CO_2 Used in EAF Process

Electric arc furnace (EAF) steel takes up around 10.1% of the steel production in China in 2012, which is far lower than the international average and it is related to the lack of scrap and energy in China. However, with the development of the country, the ratio of scrap is expected to increase and the EAF process would be comparable to developed countries. The use of CO_2 in EAF would have a great impact in future steel production in China. CO_2 was introduced as the bottom blowing gas instead of Ar in the EAF process, in order to enhance the stirring ability of the bath for promoting the slag-metal reaction, homogenize the bath temperature and composition, and increase the yield of valuable metal (Gulyaev & Filippov, 2001; He et al., 1992; Van Wijngaar den & Pieterse, 1994; Zhao, 1995). The materials and heat balances were calculated by Wang et al. (2012) for the introduction of CO_2 in EAF process with hot metal charging.

These authors concluded that, with the CO_2 addition in the oxidization gas, the hot metal ratio charged into EAF could be enhanced, resulting in energy savings. Recently, the industrial experiments for CO_2 application in the EAF process were carried out in a 65t EAF with the product of 45# steel (Wang et al., 2014) compared to Ar bottom stirring. The results show that with CO_2 bottom blowing, the stirring effect was reinforced and the FeO content in slag was lowered, which offers good desulphurization conditions. The desulphurization ratio could be increased by 7% compared to that with Ar bottom blowing. However, with CO_2 bottom blowing, the end point carbon content will increase slightly, and a small amount of Cr will be oxidized because CO_2 oxidizes Cr in the melt, while Ar has no effect on Cr oxidation.

To conclude, with CO_2 bottom blowing in EAF instead of Ar, the production cost will be decreased because the expensive Ar is replaced and the quality of the steel will not be affected, even enhanced at times.

3.3.4 CO_2 Used in Secondary Refining Process

The use of CO_2 as protection gas in a ladle furnace (LF) during heating can be traced back to 1986, with the aim to form a CO_2 gas layer. This could reduce the possibility of steel melt absorbing nitrogen as well as re-oxidation. American and French plants have applied this technology successfully to decease the N content in the liquid steel by 63–75% (Hara et al., 1986). The research work by Bruce et al. (1987a) suggested that the quality of molten steel is not affected by bottom blowing CO_2 in LF operation.

Unfortunately, because of weak oxidation of CO_2, few researches are focused on applying CO_2 as a bottom stirring gas in LF operation. Because of this background, the current group, Gu et al. (2014) and Dong et al. (2014) have studied the CO_2 application in LF both experimentally and theoretically. The experiments were performed in a 70 ton LF with CO_2 as bottom blowing gas for 45# grade steel production. The results showed that neither types nor compositions of inclusions have significantly changed in steel compared with bottom blowing Ar.

On the other hand, it is observed that the process decreases inclusion yield density and improves steel cleanness. But one should also take note of the fact that CO_2 is a weak oxidizer, and its use may lead to the increase of oxygen content in the liquid steel. A slight increase of oxygen in steel has been observed by Gu et al. (2014) and Dong et al. (2014) when blowing 1/3 and 2/3 CO_2 compared with pure Ar blowing.

Further evidence is obtained by experiments conducted by Paules et al. (1987) and Bruce et al. (1987b), who demonstrated that, when the carbon content is less than 0.20wt% in the melt, the dissolved oxygen in liquid steel is higher with CO_2 stirring than with Ar stirring, but for steel grades with higher carbon content, there is no difference of dissolved oxygen content by blowing these two gases.

3.4 CO_2 Used in Continuous Casting Process

CO_2 was applied in casting as shield gas, replacing Ar for special steel rod production in America 1989 (Anderson et al., 1989). Almost at the same period, researchers from UK (Normanton & Wang 1990) also tested the utilization of CO_2 on ladle slide gate as shield gas during 1986–1988. The conclusion from these experiences was that, with the protection of CO_2, the steel quality was similar to that when Ar is used. The utilization of CO_2 in the casting process came into popularity rather late in China. Until 2015, Li et al. (2015b) have carried out the industrial experiments on CO_2 applied as a protective gas during continuous casting process instead of Ar/N_2.

When CO_2 is used as stirring gas or shielding gas for production, the quality of steel is similar or even the same to that using Ar gas. The nitrogen content in the melt is slightly lower with CO_2 blowing than with Ar blowing since 1 volume CO_2 reacting with C will generate 2 volumes of CO and large amounts of gas benefit from removing nitrogen. The oxygen content, hydrogen content and the content of oxide inclusions in the liquid steel are comparable with that stirred by Ar.

3.5 CO_2 Used in Ferroalloy Production

Converter refining for ferrochrome (FeCr) and ferro manganese (FeMn) production is the trend of medium and low carbon (MLC) product smelting in China. Because the submerged arc furnace used for high carbon ferroalloy production is getting larger in size and the tapping size is very large. The conventional way for medium and low carbon ferroalloy production is not suitable because of its limited size and very high energy consumption. But the current converter technology used for M-LCFeCr and M-LCFeMn production has defects of a short lifetime of linings and low yield of Cr/Mn. In this case, current authors (Yu, 2015; Yu et al., 2015) introduced CO_2 into the converter process for M-LCFeCr and M-LCFeMn production. Laboratory scale experiments show that it is feasible to use CO_2 for producing these two alloys and it can enhance the yield of Cr/Mn by using $CO_2 + O_2$ mixtures, especially when the carbon content in product is higher than 3.0wt%. However, calculations indicate that it is not beneficial to produce ferrochrome with carbon content lower than 1.0wt% with the introduction of CO_2.

Additionally, the refining of silicon and ferrosilicon with the outfurnace method has also considered using CO_2 as a medium; the results, however, are not very encouraging (Suzuki et al., 1992; Wu et al., 2009, 2014) and would require further study.

3.6 CO_2 Used in Other Areas in Steel Process

CO_2 was also used as the protection media, as a substitute for hydrocarbon, for side-blown oxygen lance in a converter to successfully limit dissolved hydrogen in liquid steel (Liu, 1983). As an acidic gas, CO_2 could also be used to fix active basic substances in slag to get stable carbonate. Dong (2008) have used CO_2 to stabilize the converter slag and found that the best reaction temperature was 700 °C. Slag splashing in a converter is widely used in China and CO_2 is being tested as a replacement for N_2 by researchers at the Chinese Academy of Sciences (Yuan et al., 2007; Zhao et al., 2010). During these trials with slag splashing, coke fines or pulverized coal with additives were added into the converter. The basic idea was that CO_2 would react with C to form CO. In this case, $CO + CO_2$ mixed gas will be recycled using this process. After separation, the CO_2 could be utilized for slag splashing again, while CO could be reused in other places in plant.

3.7 CO_2 Used in Microalgae Cultivation

CO_2 used in microalgae cultivation is another promising way of sequestering CO_2. On one hand, microalgae could fix carbon in a very efficient way with zero carbon emissions. On the other hand, it helps to realize the transformation of solar energy to biomass energy. Microalgae has broad applications. It can be used to produce biodiesel which has many advantages over traditional fuels. Advantages include better combustion and safety behavior, and higher environmental performance. Another microalgae application is in food industry. Microalgae is used to produce products with high industrial added value, like docosahexaenoic acid and eicosapentaenoic acid.

Biofuels and methanol, which are emerging class of carbon–neutral drop-in replacement fuels, offer strongly viable alternatives for providing carbon neutral fuels to large and very large vessels. At the moment, through biotechnology utilization creates a carbon neutral loop as the CO_2 released from burning the biofuels and methanol are captured and offset from industrial activities in the production of biofuels and methanol. Energy need for the reaction is in the renewable form. Utilization biological carbon capture there will convert 70,000 t per year of CO_2 into 50,000 t per year of methanol (ADB, 2020).

The broad geographical distribution of microalgae and its extraordinary environmental tolerance make large-scale microalgae cultivation possible. China is the biggest microalgae breeding country worldwide and has great potential for microalgae development. Therefore, microalgae technology should be perceived as an important development direction of carbon fixation and should receive more attention. Currently, technical challenges include the low quality of microalgae, less efficiency of large-scale cultivation, and high energy cost.

3.8 Econic Catalyst Technology for Utilization of Carbon Dioxide as a Feedstock for the Polyurethane Industry

The proprietary catalyst technology of Econic allows for recycled carbon dioxide (CO_2) emissions to be used as a feedstock in polyol manufacture. In the frst market application, recycled CO_2 is incorporated into a range of polyols—the building blocks of polyurethanes—substituting up to 50% of the conventional fossil-based feedstocks, propylene oxide and/or ethylene oxide. Econic technology produces polyols with tunable amounts of incorporated CO_2, which have enhanced performances that are tailored for specifc applications. Beyond polyols,

Econic technology has been proven to deliver the same CO_2 benefts in the surfactant, lubricant, and high molecular weight polymer sectors. Its frst target market is in the manufacture of polyols, the building blocks of polyurethane materials. Life-cycle assessment studies indicate that polyols produced with average CO_2 contents (25%–35%) will have reduced CO_2 emissions footprints in the order of 25–40%. Across the global polyurethanes market, a 30% adoption of Econic catalyst technology could result in the consumption and prevention of 5 million tonnes per annum of CO_2 emissions. Polyols created by the Econic system can be utilized in all aspects of the polyurethane market, from housing insulation to car interiors to running shoes and refrigerators. The Econic system incorporates CO_2 into polyols using less energy than the existing industrial process, and—compared to competing technologies—is the only system that facilitates bespoke incorporation of CO_2 into products, thereby optimizing production and end-product performance. Furthermore, it does so under the same equipment and processing conditions of existing production facilities, allowing fast, simple, and low-cost adoption by retrofit.

Beyond polyols, the same utilization of CO_2 has been demonstrated in the surfactant, lubricant, and high molecular weight polymer sectors, dramatically increasing the scale of carbon beneft the Econic technology can deliver.

3.9 Carbon Utilization Recommendation

Significant research efforts have been made on CO_2 used as stirring gas, reacting medium as well as protecting source in metallurgical processes. This covers almost all the fields of ferrous metallurgy. In most cases, positive results are obtained through observation. However, despite the positive impacts, the negative effects of CO_2 also need to be considered. For example, when it is used in converter and EAF processes, it may prolong the tap-to-tap time hence increase the production cost somehow, and it is difficult to decide the proper introduction amount. While, when it is used in LF (ladle furnace) and CC (continuous casting) processes, it may lead to re-oxidation of the melt for clean steel production.

Till now, the experiments on application of CO_2 in the metallurgical processes in China are limited to laboratory scale and pilot scale. The lack of enthusiasm in industrial implementation could be attributed to (1) method of CO_2 preparation leads to the cost of CO_2 currently is not low, and it is not economical to use CO_2; (2) there is lack of sufficient research results with regard to the optimal amount of CO_2 to be used. These problems need to be solved by researchers as soon as possible. Recently, the new methods for CO_2 separation from the flue gas in steel industry reported by researchers from South Korea (Kim et al., 2015) provided an economic solution of CO_2 preparation (62 USD/t CO_2 separated), which will make CO_2 cyclic utilization in the metallurgical processes closer to the industrial production.

4 Review of Appropriate CO_2 Transport and Storage Options

4.1 Options for CO_2 Transport

Generally, a CCS system has three separate elements: capture, transport and storage. As the intermediate process in the CCS chain, CO_2 transport is needed to link CO_2 sources and storage sites and thus plays a significant role in the successful deployment of CCS. Hence, it is important to have a thorough consideration of CO_2 transportation approaches and corresponding cost. However, transport is considered as the least challenging element of CCS and has been less studied (Aspelund & Jordal, 2007).

In industry, CO_2 is transported by pipelines, ships, railway tankers, and tanker trucks. Svensson et al. (2004) performed an economic analysis on these four alternatives and concluded that water carriers (offshore), pipelines (offshore and onshore) and a combination of these are appropriate transport options for CCS. Figure 2 shows the cost and capacity for transportation alternatives at a distance of 250 km. It clearly shows that pipeline and water carriers have a large capacity and low cost. Nevertheless, other options still have the potential to be applied to small-scale CCS projects. For example, Yanchang integrated a CCS demonstration project in China that adopts tanker trucks and pipelines as its CO_2 transport

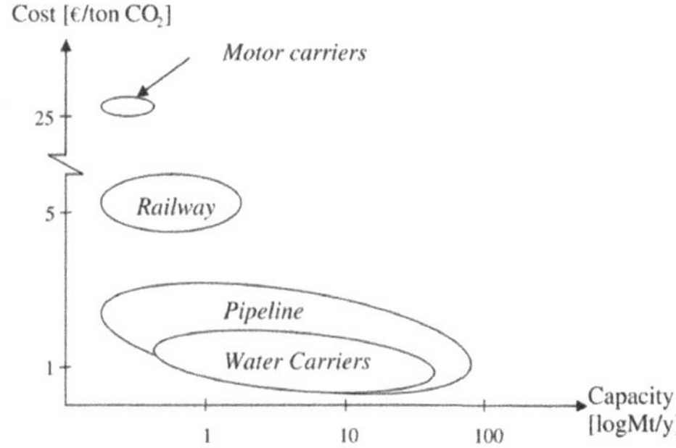

Fig. 2 Cost and capacity for transportation alternatives at 250 km (Svensson et al., 2004)

method (GCCSI, 2017a). For large-scale CCS projects, due to the low capacity and high transportation fares, railway tankers and tanker trucks are not cost effective. Therefore, this portion of the review will focus on pipeline and shipment transport and gives a brief introduction of transport technology as well as cost analysis.

4.1.1 CO_2 Pipeline Transport

The use of CO_2 pipelines is well established to be the most economical way of transporting CO_2 (Zhang et al., 2006). At present, 14 of 16 global large-scale CCS projects use pipeline transport (Table 1). Carbon dioxide pipelines have extended over 2500 km in the western areas of United States, where they transport more than 50 million tonnes of CO_2 per year from natural CO_2 sources to enhanced oil recovery sites. By far, many problems associated with carbon dioxide pipelines are similar with those encountered by hydrocarbon pipelines in similar areas. Therefore, lots of learning and knowledge from the use of hydrocarbon pipelines, i.e. system design, construction, risk monitoring, can be borrowed to develop pipelines for CO_2 transport.

Transport Technology and System Design

Broadly speaking, the CO_2 pipeline transport chain includes gas conditioning, compression and pipeline transport (Fig. 3). Gas conditioning means the necessary processing of CO_2 to achieve transport specification. Normally, it includes two steps: removing water and other liquids and then removing volatile gases. Liquids should be ensured not to enter CO_2 compressors so that the formation of gas hydrates, the freezing of water and corrosion can be avoided. Transportation of volatile gases, like nitrogen, would increase the energy requirement and cost. Gas conditioning is often recognized as the interface between CO_2 capture and transport.

The most economical way of transporting CO_2 is as a dense phase (Vandeginste & Piessens, 2008). The dense phase pipeline was 20% cheaper than a low-pressure CO_2 gas pipeline. Generally, the pressure should be kept above 9.6 MPa to remain in a dense phase state no matter the temperature change. In addition, high pressure CO_2 is advantageous in the injection process. Compressors not only offer the pressure at the starting point, but also operate at some points along the pipeline, at areas called booster stations, to overcome the pressure loss caused by friction. These stations are not necessarily equally spaced. When pipeline length is short, they may lack compressor stations.

Table 1 Transportation of global large-scale CCS project in operation. Adapted from GCCSI (2017b)

Project name	Capacity (Mtpa)	Operation date	Transport type	Length (km)
Val Verde Natural Gas Plants	1.3	1972	Pipeline	316
Enid Fertilizer CO_2-EOR Project	0.7	1982	Pipeline	225
Shute Creek Gas Processing Facility	7.0	1986	Pipeline	Multiple, maximum of 460 km
Sleipner CO_2 Storage Project	0.85	1996	N/A	Not applicable
Great Plains Synfuel Plant and Weyburn-Midale Project	3.0	2000	Pipeline	329
Snøhvit CO_2 Storage Project	0.7	2008	Pipeline	153
Century Plant	8.4	2010	Pipeline	64 to 240
Air Products Steam Methane Reformer EOR Project	1.0	2013	Pipeline	158
Coffeyville Gasification Plant	1.0	2013	Pipeline	112
Lost Cabin Gas Plant	0.9	2013	Pipeline	374
Petrobras Santos Basin Pre-Salt Oil Field CCS Project	Approx. 1.0	2013	N/A	Not applicable
Boundary Dam Carbon Capture and Storage Project	1.0	2014	Pipeline	66
Uthmaniyah CO_2-EOR Demonstration Project	0.8	2015	Pipeline	85
Quest	Approx. 1.0	2015	Pipeline	64
Abu Dhabi CCS Project (Phase 1 being Emirates Steel Industries (ESI) CCS Project)	0.8	2016	Pipeline	43
Petra Nova Carbon Capture Project	1.4	2017	Pipeline	132

N/A means no transport required (direct injection)

The diameter of the pipeline is the crucial parameter in pipeline design. Based on the mechanical energy balance, a model was developed to calculate the pipeline diameter required for a given pressure drop (Mohitpour et al., 2003). The diameter is also influenced by mass flow rate, fanning friction factor and pipeline length. The solving process is iterative, so an initial guess is needed.

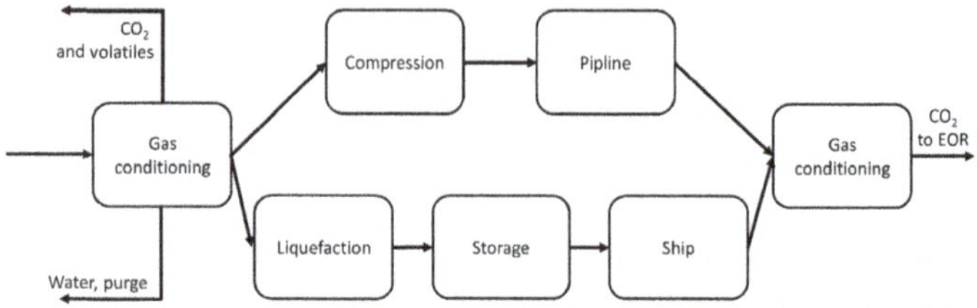

Fig. 3 The CO_2 transport chain (Modified from Aspelund & Jordal, 2007)

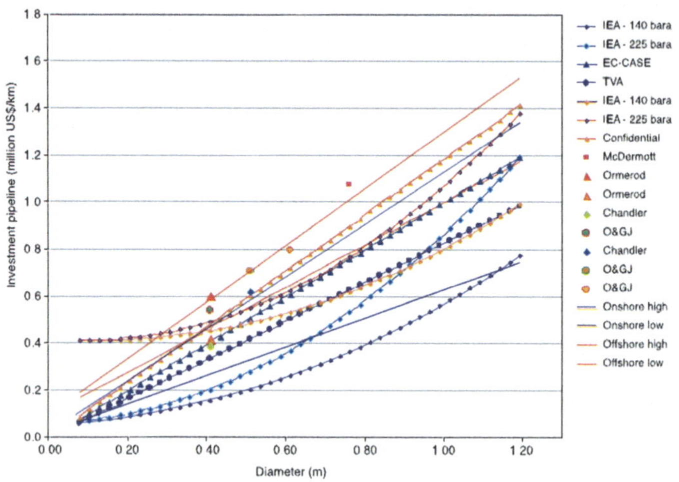

Fig. 4 Construction costs for pipelines from various information sources. Costs exclude booster stations (Metz, 2005)

A monitoring system is essential as high concentrations of CO_2 can be fatal for humans and animals. Pipelines can be monitored internally by pigs, which are pressure-driven devices with piston shapes moved along the pipeline, and can be monitored externally by leak detection systems.

Economic Analysis of Pipeline Transport

The cost of CO_2 transport includes three parts. Construction costs, also called capital costs, measure the initial investment costs. It contains material/equipment costs, and labor costs. Figure 4 shows construction costs for pipelines in different contexts. This data clearly shows the relationship between diameter and the capital cost. It is also obvious that booster stations increase the capital cost. Operation and maintenance costs include monitoring costs, maintenance costs and energy costs. Other costs are miscellaneous, including design, project management and right-of-way costs, etc. Levelized cost takes all the costs in the whole lifetime of a project into account (Fig. 5), and it represents the cost of per tonne CO_2 transported by pipeline (McCollum & Ogden, 2006).

Based on natural gas pipelines experience, Mccoy and Rubin (2008) developed a model for prediction of the pipeline capital expense in United States. For an illustrative example (transport of five million tonnes of CO_2 with distance larger than 100 km), capital cost was US$1.16 per tonne of CO_2. Liu and Gallagher (2011) developed a model which considers both engineering and economic factors with the background of China. For five illustrative case studies with a pipeline length of 100 km, the levelized costs are US$1.84–3.06 per tonne of CO_2.

4.1.2 CO_2 Transport by Ship

CO_2 transport by ship is more flexible compared to pipeline transport, and it is more competitive for ocean CO_2 storage or marine oilfield EOR. CO_2 transport by ship is very similar to liquefied petroleum gas (LPG) transport by ship. However, the number of CO_2 marine tankers is rather limited because of the low demand. Currently, there are no existing experienced CO_2 transport ships for CCS. Only four small-scale semi-pressurized CO_2 ships operate in western Europe, the CO_2 of which is used for the food industry. Korean-CCS 1 and 2 may become the first two projects using shipment transport for CO_2, as the storage site is offshore Korea. Now they are under evaluation and the operation date is in the 2020s.

Transport Technology and System Design

CO_2 transport process by ship is quite different compared to the process by pipeline. The ship transport chain of CO_2 includes liquefaction and gas conditioning, an intermediate storage, a loading system, shipping and a loading system. At the

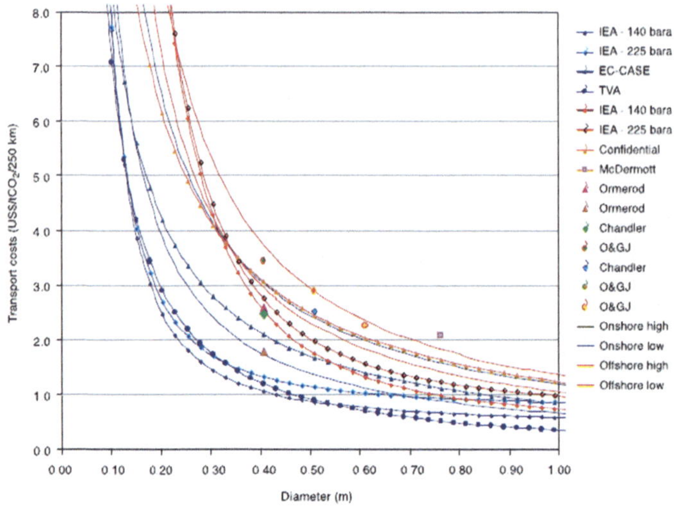

Fig. 5 Levelized cost of pipeline transport from various information sources (Metz, 2005)

onset of the transport process, CO_2 is liquefied around its triple point. The density of CO_2 is as high as 1200 kg/m^3 at its triple point and reduces to half at its critical point. Higher density means more CO_2 can be transported for the same size ship. In addition, since CO_2 is captured continuously but transported separately, the system must include an intermediate storage part. The CO_2 is contained in semi-pressurized tanks before being loaded on board.

There are three types of ships for transporting CO_2: fully pressurized ship, semi-pressurized ship and fully refrigerated ship. Semi-pressurized ships are recommended by ship designers and the best transport condition is at 0.65 MPa and −52 °C (Bock et al., 2003). LPG ships can be rebuilt for CO_2 transport. However, some difficulties may arise because the CO_2 is transported at a little higher pressure and lower temperatures. It is noteworthy that CO_2 leakage during ship transport is inevitable. The total loss is estimated between 3 and 4% per 1000 km (Metz, 2005).

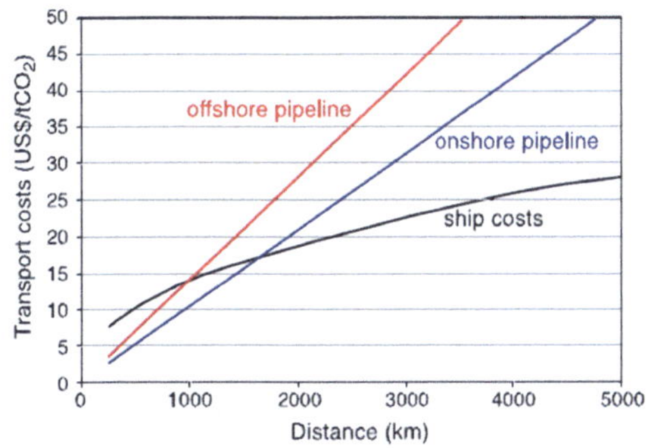

Fig. 6 Comparison of costs of onshore and offshore pipelines, and ship transport (Metz, 2005)

Economic Analysis of CO_2 Transport by Ship

Costs of shipment system are composed of capital costs, operational and maintenance costs and miscellaneous costs. However, less is known on the costs of a CO_2 shipment system since no large-scale CCS projects adopt this means of transportation by far. Therefore, the cost estimates from different literature vary widely.

Aspelund et al. (2006) performed a cost, energy and CO_2 emission analysis on transport chain within the North Sea. The total cost is US$20–30 per tonne of CO_2 for volumes larger than 2 Mt per year. The energy requirement is 142 KWh per tonne of CO_2, and the liquefaction process accounts for the major energy consumption with a proportion of 77%. Total CO_2 emissions are estimated to be 1.4% of the inlet CO_2. By comparison, a case study in China gives the levelized cost of shipment transport as US$7.09 per tonne of CO_2 with a ship transport distance of 300 km and pipeline length of 25 km (Gao et al., 2011). The ship capacity and travel distance are sensitive factors and will have large impact on cost. The relationship between travel distance and the levelized cost is non-linear. Therefore, CO_2 transport by ship is a more economical solution when the distance is greater than 1000 km (Aspelund et al., 2006). Figure 6 clearly shows as distance increases, ship will be the cheapest way to transport CO_2.

4.1.3 Proximity Study

Broadly speaking, any place that releases CO_2 into the atmosphere is a CO_2 source. In contrast, any place that sequestrates CO_2 from the atmosphere is a CO_2 sink. However, in this context sources and sinks refer to those related to CCS project. Therefore, CO_2 sources include power plants, steel mills, cement factories, etc. CO_2 sinks include oil reservoirs, deep saline aquifers, coal beds and the ocean.

Proximity study is indispensable in CCS system. It is necessary to spot the geo-location of CO_2 sources and sinks and specify their capacity so as to evaluate the CCS potential in an area. In addition, the geological information can be used to calculate the transportation cost. The availability of CO_2 sources data is relatively easy, but things are not easy for sinks. The sinks are always underground which means geological exploration method should be used. In this case, the accuracy is compromised. In most cases, CO_2 sources and sinks do not coincide. So proximity study is needed to deploy CCS transport lines wisely to optimize the NPV.

Proximity study of China is at an early stage, though several studies have estimated CCS potential of China. Bai et al. (2006) performed a research on CO_2 point sources in industry and studied how they distribute across China. Liu et al. (2005) made estimation on CO_2 storage capacity of coalbeds in China. Li et al. (2006) studied domestic deep saline aquifers and ranked their potential for CCS application. At present, proximity study of China is at national scale which is not meaningful considering a specific CCS project. Also the lack of integrated study of sources and sinks undermines the integrity of CCS system. Therefore, proximity should be performed at regional scale and be more holistic in future.

4.1.4 CO_2 Transport Recommendation

CO_2 transport approaches should be determined based on proximity study and the quantity of CO_2 to be sequestrated. For the existing alternatives, pipeline transport is the most economical and stable method for terrestrial storage. However, for some special cases, i.e. CO_2 capture from marine natural gas production, ship transport can be an alternative option. In summary, pipeline transport is our recommendation for CO_2 transport. We should focus on building of pipeline networks.

4.2 Trapping Mechanisms of Geological Storage of CO_2

Trapping mechanisms of CO_2 can be classified into five types: structural, capillary, solubility, mineral and hydrodynamic trapping (Fig. 7). Contribution of different mechanisms is dynamic and the dominant mechanism is changing at different phases as CO_2 migrates and reacts with rocks and fluids (Benson & Cole, 2008). With the extension of time, the storage security of CO_2 sequestration increases. The trapping mechanism determines the contribution of each trapping approach. At the beginning, structural trapping and hydrodynamic trapping play an important role. After hundreds of years, the effect of capillary, solubility and

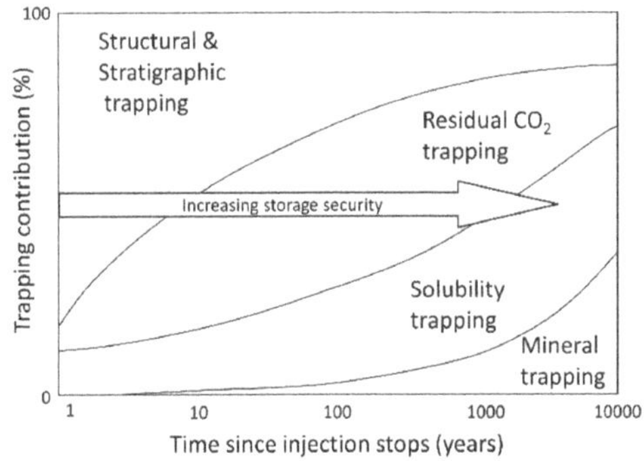

Fig. 7 Trapping mechanisms evolution with time (Modified from Metz et al., 2005)

mineral trapping increases in importance. All the mechanisms will be introduced in detail in following sections.

For scientific research on caprock-sealing mechanisms of CO_2 sequestration, Song and Zhang (2013) presented a comprehensive overview of the CO_2 caprock-sealing mechanisms, where the problems related to CO_2 leakage are defined, and major research gaps and additional study areas are identified.

4.2.1 Structural Trapping

Structural trapping is of great significance in early stages of CO_2 injection. A thick and fine-textured rock, such as shale, can serve as a caprock above the sequestration reservoir and prevent CO_2 from migrating freely. Though the buoyance of CO_2 is large, the block effect of the impermeable layer impedes the vertical migration of CO_2. Furthermore, the caprock could prevent CO_2 from migrating upwards due to its low permeability and high capillary entry pressure (Benson & Cole, 2008). The geological structure apt to such trap includes anticline, fault block, and pinch. In most cases, the structure originally contains oil and gas, or groundwater.

4.2.2 Capillary Trapping

After injection when CO_2 migrates into the formation, CO_2 can be immobilized in pores within rocks because of the gas–liquid interface tension (IFT), which is called residual trapping. When large quantities of CO_2 pass across porous media, CO_2 is usually trapped in the pore as spherical drops. Therefore, the more the amount of CO_2 passing the porous media, the more the CO_2 is trapped in the rock pore. Normally, this process can last for several decades from the time CO_2 once being injected into the formation.

4.2.3 Solubility Trapping

Another trapping mechanism, solubility trapping, starts to function when CO_2 migrates in the pore of rocks. CO_2 will contact with water and crude oil in which CO_2 can dissolve. The solubility of CO_2 is influenced by several elements, including temperature, pressure and fluid salinity. Other influencing elements that could affect the solubility of CO_2 include the chemical components of formation fluids, and the contact rate between CO_2 and formation fluids. Because of the gravity, CO_2-saturated formation fluids will move towards the center of the basin over time. Therefore, solubility trapping has more storage potential compared with structural trapping. The time-scale of solubility trapping is from 100 to 1000 years.

4.2.4 Mineral Trapping

Mineral trapping mechanism happens once CO_2, either in dissolved state or supercritical state, reacts in a direct or indirect way with formation rocks, which in most cases promotes carbonate mineral precipitation (Oelkers et al., 2008). Although mineral trapping is considered as the most reliable for the reason that it can last for very long periods, the reaction happens very slowly. The overall efficiency is usually achieved in a timescale of hundreds of years, or maybe longer. The main factors which can influence the speed of mineral trapping include chemical reaction process, fluid type and mineral composition. For example, the chemical reaction rate is high when the formation is carbonate. In contrast, the chemical reaction rate is low or even no reaction happens when the formation is sandstone.

Besides geologic formation, concrete may be another choice when considering mineral trapping. A novel technology from CarbonCure injects CO_2 into fresh concrete during production (ADB, 2020). The injected CO_2 has been proved to strengthen concrete mechanical properties, which can reduce cement requirement and optimize the mix designs. Most importantly, CO_2 can be permanently trapped in concrete due to mineralization. This technology provides a new approach to storing and reusing waste CO_2.

4.2.5 Hydrodynamic Trapping

The groundwater flow pressure will impede and accumulate CO_2 when its direction is opposite to that of CO_2 buoyancy and of approximately equal force. This mechanism works when the CO_2 is injected into the deep saline aquifers within a closed formation. Like the structural trapping, hydrodynamic trapping works immediately after the injection. The difference is that the lateral movement of CO_2 is not blocked in hydrodynamic trapping. The groundwater migration velocity is measured in centimeters per year, whereas the CO_2 migration distance is measured in kilometers. So it will take thousands of years for the injected CO_2 to migrate to the shallower geologic sections. Other trapping mechanisms will likely also contribute and, consequently, no free CO_2 will reach the surface. Thus the integrity of the CO_2 storage is guaranteed.

4.3 Options for CO_2 Storage

Generally speaking, optimal CO_2 storage sites need to be located at the edge or inside a stable plate, such as a craton basin. The size and effectiveness of the basin storage space directly determine the amount of CO_2 stored, and finally, the pressure and fluid system of the formation water in the basin are also crucial. important. According to the current research level, three basin geological bodies are appropriate to CO_2 underground burial: depleted oil and gas reservoirs, saline aquifers, and coal seams (Fig. 8). Each structure type has a certain storage capacity, but due to different storage media, its storage conditions also have certain differences. Different options for CO_2 storage have their advantages and disadvantage as discussed below.

4.3.1 Depleted Oil and Gas Reservoirs

Oil and gas reservoirs, especially depleted oil and gas reservoirs, are preferred for underground geological storage of CO_2 due to following advantages: (1) Good trapping conditions (structural and stratigraphic), gas can be sealed for a long time with a small risk of leakage; (2) High degree of exploration, including the

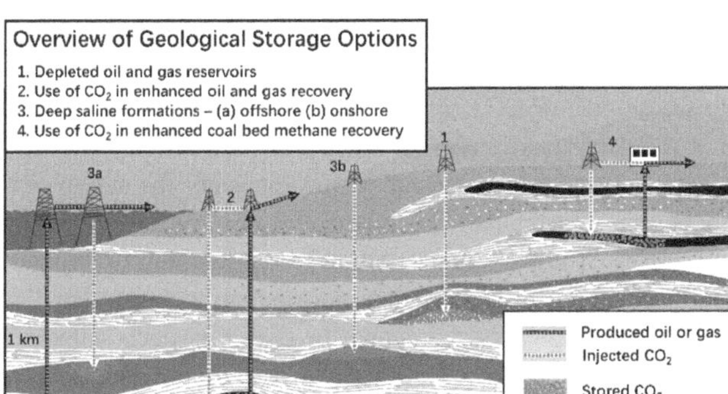

Fig. 8 Different kinds of appropriate CO_2 storage sites (Modified from Benson & Cole, 2008)

geological structure and physical properties; (3) Well-developed computer simulation technology and models for the prediction of trapping and displacement of hydrocarbons; (4) More than 80% of oil and gas reservoirs lie in deep formations. Pressure and temperature conditions are relatively easier to store CO_2 in the supercritical state; (5) Complete infrastructure and wells constructions.

These conditions are the main- criteria for measuring whether a depleted oil field is capable of underground CO_2 storage. However, care should be taken to avoid exceeding the pressure that the caprock can withstand, and the reservoir is not easily affected by permeability reduction induced by plugging near injection well and fluctuations in reservoir stress, which will cause the reservoir capacity to face certain limitations (Bossie-Codreanu et al., 2003; Kovscek, 2002). In addition, CO_2 injection can also increase oil recovery rate, enhance oil and gas production and protect environment.

4.3.2 Saline Aquifers

The underground saline aquifer suitable for geological storage of CO_2 generally refers to a rock formation with high porosity and permeability that is filled with formation waters or brines, and the salt concentration of which is pretty high at a certain depth (at least 800 m) (Gunter et al., 1993). These filled salt waters (with a mass concentration of NaCl of 32×10^{-3} mg/L) are unsuitable for industrial or agricultural use. These saline aquifers can be found in most sedimentary basins in the world. Around 2 percent of the available space can be used to storage CO_2 (Metz et al., 2005). The reservoir property directly affects the effectiveness, safety, and durability of storage.

The stratum depth of the top interface of the saline aquifer must be at least 800 m or more to guarantee the supercritical state of CO_2. If the top and bottom of the saline aquifer are blocked or enclosed by shale or gypsum-salt layers, the formation water will have a high salinity due to slow flow, which will be very conducive to the storage of CO_2, which can stay for millions of years.

Because of the density difference between CO_2 and formation water, the CO_2 injected into the saltwater layer will move upward under the action of buoyancy. This movement appears irregular due to being blocked by low permeability rock formations or lenses. If it encounters a suitable stratigraphic or structural trap, it will be stored; while if there are no such blocking or enclosure conditions, CO_2 may continue to move upwards and cannot be buried in deep saline aquifers. In addition, in general sandstone or carbonate rock, pores or cracks are relatively developed, and most of them are easily filled by water to form saline aquifers. On the contrary, crystalline and metamorphic rocks, like granite, are not applicable to the formation of saline aquifers and therefore to CO_2 storage.

The CO_2 injected into the stable saltwater layer is mainly dissolved in the early stage of the chemical reaction. Under normal circumstances, the solubility of CO_2 in saline aquifers will increase with pressure, temperature, and salinity, and it is also related to the contact with formation water. Experiments show that under certain salinity and depth conditions, assuming that the saline aquifer is more homogeneous and given sufficient dissolution time, 1 m^3 of formation water can dissolve about 20–60 kg of CO_2; if the injection continues for 25 years, then 10–25% of CO_2 will be dissolved by formation water (Gale, 2003).

4.3.3 Coal Seams

In coal beds, the storage of CO_2 is mainly realized by the CO_2 adsorption process. Injected plume of CO_2 is mainly within the cleat between the coal blocks and inside the coal block (Rice et al., 1993). There are many tiny pores developed in the coal block sandwiched between the endogenous fissures, which can fully and firmly absorb the gas molecules from the fissures, thus the endogenous fissures between the coal blocks are important in the coal seam. The pore system is very important in the process of gas transport between different seams. Therefore, this transport effect directly affects CO_2 injection and storage prospects. At present, the majority of coal seams worldwide suffer from low permeability, the average of which is only $1\text{–}5 \times 10^{-3}\ \mu m^2$. In addition, the formation thickness is relatively thin, the average is only 1–5 m or part of >5 m (Larsen, 2004). This unfavorable condition directly restricts the storage capacity of CO_2 in coal seams and specific storage projects implementation.

4.3.4 Deep-Sea Sediments

Although the above-mentioned storage method has good application prospects, there is still a risk of leakage because of the pressure rise induced by CO_2. Scientists from many countries have begun to pay attention to deep-sea sediments.

The latest research (Teng & Zhang, 2018) studied the long-term evolution process of carbon dioxide injected into the seabed sediments, carried out numerical studies in a computer, and evaluated the effect of storage. Studies have found that in a high-pressure and low-temperature environment, the high density of liquid carbon dioxide will cause "negative buoyancy" in a sinking state. Coupled with the reaction of carbon dioxide with water to form carbon dioxide hydrate and block the flow channel, it can affect the carbon dioxide in the formation. The floating situation plays a very good role in hindering, so as to realize the safe, stable, and permanent storage of CO_2. On the other hand, the CO_2 stored by this method will not come into contact with seawater, which avoids the negative impact on the marine ecological environment that may be caused by conventional deep-sea CO_2 storage.

This research provides a novel solution for geological storage of CO_2 in addition to land, expanding the storage sites to a wide range of marine sediments, and has important significance and application prospects for the future global carbon dioxide emissions reduction and storage options.

Since the operating facilities required for deep-sea geological storage are similar to the semi-submersible offshore platforms designed for combustible ice mining in the sea, and the injection of carbon dioxide is conducive to the mining of combustible ice, the two can be carried out simultaneously. The next step of research may try to actively seek opportunities to combine with field experiments on the combustible ice mining.

4.3.5 CO_2 Storage Recommendation

In light of the large quantities of coalbeds and oilfields in China, ECBM and EOR are recommended. On the one hand, the CO_2 storage potential is large. On the other hand, the increase in coal gas and oil production could be profitable. It is also necessary to improve the techniques for CO_2 storage to lower the cost and leakage risk. For example, a new monitoring system is needed. Finally, we should attract more and more enterprises to participate in the CCS projects. It is very beneficial to expand the scale of CCS and increase the public awareness.

4.4 Case Studies

4.4.1 CCS Projects in the World

As the world's first commercial CO_2 storage project, Sleipner operated at 1996 and millions of tonnes of CO_2 were injected underground in the middle of the Central North Sea ever since. Covering an area of about 26,000 km^2, the shallow sands in Utsira Formation act as an ideal site for CO_2 storage. The thickness of the formation is between 200 to 300 m and porosity is up to 40% and permeability is above 1 Darcy. High porosity and large permeability enable Utsira Formation to storage large quantities of CO_2. The estimated storing capability of Utsira Formation is up to 600 billion tonnes.

From Table 2, we can see that most of world's massive CCS projects utilize CO_2 to increase oil production rate. The other three projects inject CO_2 into deep saline aquifers for storage.

4.4.2 CCS Projects in China

Many Chinese energy-related companies have made efforts on CCS development and a series of CCS projects and many demonstration CCS projects have been implemented. In 2011, Shenhua Group picked Ordos Basin to demonstrate their CCS project and began to sequestrate CO_2 in saline aquifers. Two largest oil companies in China, CNPC and Sinopec, have launched numerous CO_2-EOR operations in many plays, and many academic achievements are made (Wu et al., 2014). Huaneng Group has also attempted to build devices to collect CO_2 from flue gas in large Chinese cities. China United Coalbed Methane Co., Ltd. has been carrying out CO_2-ECBM projects in two wells. The CCS project in Yanchang, Shanxi Province has also come into the implementation stage. Every year, around half a million tonnes of CO_2 is collected from local chemical plants, and the CO_2 was then transported to the storage site.

4.4.3 The Next Wave of Projects

The Paris Agreement sets a plan of keeping global temperature increase below 2 °C by 2050, and CCS plays a key role for achieving this target. However, the overall trend of CCS is not promising. Around two thirds of CCS projects are unable to reach a financial investment decision. By 2016 the large-scale CCS projects had shrunk to 38 from 77 large-scale CCS projects in 2010. It is clear to see that the pace of global CCS project development is stagnated and even goes backwards. (IEA, 2016) concludes that CCS is off track for Paris Agreement goals and picking up the pace of CCS deployment is urgent. Accelerating the pace of CCS deployment needs more policy support, economic investment, advanced technology, and global cooperation.

Table 2 Primary capture and storage type of global large-scale CCS project in operation. Adapted from GCCSI (2017b)

Project name	Country	Operation date	Industry	Capture type	Primary storage type
Val Verde Natural Gas Plants	United States	1972	Natural gas processing	Pre-combustion capture	Enhanced oil recovery
Enid Fertilizer CO_2-EOR Project	United States	1982	Fertiliser production	Industrial separation	Enhanced oil recovery
Shute Creek Gas Processing Facility	United States	1986	Natural gas processing	Pre-combustion capture	Enhanced oil recovery
Sleipner CO_2 Storage Project	NORWAY	1996	Natural gas processing	Pre-combustion capture	offshore deep saline formations
Great Plains Synfuel Plant and Weyburn-Midale Project	CANADA	2000	Synthetic natural gas	Pre-combustion capture	Enhanced oil recovery
Snøhvit CO_2 Storage Project	NORWAY	2008	Natural gas processing	Pre-combustion capture	offshore deep saline formations
Century Plant	United States	2010	Natural gas processing	Pre-combustion capture	Enhanced oil recovery
Air Products Steam Methane Reformer EOR Project	United States	2013	Hydrogen production	Industrial separation	Enhanced oil recovery
Coffeyville Gasification Plant	United States	2013	Fertiliser production	Industrial separation	Enhanced oil recovery
Lost Cabin Gas Plant	United States	2013	Natural gas processing	Pre-combustion capture	Enhanced oil recovery

(continued)

Table 2 (continued)

Project name	Country	Operation date	Industry	Capture type	Primary storage type
Petrobras Santos Basin Pre-Salt Oil Field CCS Project	Brazil	2013	Natural gas processing	Pre-combustion capture	Enhanced oil recovery
Boundary Dam Carbon Capture and Storage Project	Canada	2014	Power generation	Post-combustion capture	Enhanced oil recovery
Uthmaniyah CO_2-EOR Demonstration Project	Saudi Arabia	2015	Natural gas processing	Pre-combustion capture	Enhanced oil recovery
Quest	Canada	2015	Hydrogen production	Industrial separation	onshore deep saline formations
Abu Dhabi CCS Project (Phase 1 being Emirates Steel Industries (ESI) CCS Project)	United Arab Emirates	2016	Iron and steel production	Industrial separation	Enhanced oil recovery
Petra Nova Carbon Capture Project	United States	2017	Power generation	Post-combustion capture	Enhanced oil recovery

Acronym	Full spelling
ADB	Asian Development Bank
AOD	Argon Oxygen Decarburization
BECCS	Bio-Energy CCS
BOF	Basic Oxygen Furnace
CC	Continuous Casting
CCS	Carbon Capture, Utilization and Storage
CO_2-EOR	CO_2 Enhanced Oil Recovery
CP	Cyclopentane
CNPC	China's National Petroleum Corporation
DAC	Direct Air Capture
DTAC	Dodecyl-trimethyl-ammonium Chloride
EAF	Electric Arc Furnace
ECBM	Enhanced Coal Bed Methane
EnOpt	Ensemble-based Optimization
EOR	Enhance Oil Recovery
ESI	Emirates Steel Industries
GCCSI	Global CCS Institute
HBCC	Hydrate-based CO_2 Capture
IEA	International Energy Agency
IPCC	Intergovernmental Panel on Climate Change
KTH	Royal Institute of Technology Sweden
LF	Ladle Furnace
LPG	Liquefied Petroleum Gas
MEA	Monoethanolamine
MLC	Medium and Low Carbon
PSA	Pressure Swing Adsorption
SDS	Sodium Dodecyl Sulphate
TBAB	Tetra-n-butyl Ammonium Bromide
THF	Tetrahydrofuran
TRT	Top-Pressure Recovery Turbine Plant
TSA	Temperature Swing Adsorption
VPSA	Vacuum Pressure Swing Adsorption

References

ADB. (2020). Carbon capture, utilization, and storage game changers in Asia.

Anderson S. H., Doulas C. L., & Bermel, C. L. (1990). Use of CO_2 in the AOD. In *47th Electric Furnace Conference Proceedings; ISS-AIME* (pp. 297–304).

Anderson, Foulard, J., & Lutgen, V. (1989). Inert gas technology for the production of low nitrogen steels. In *Electric Furnace Proceeding* (pp. 365–375).

Aspelund, A., & Jordal, K. (2007). Gas conditioning-The interface between CO_2 capture and transport. *International Journal of Greenhouse Gas Control, 1*(3), 343–354.

Aspelund, A., Molnvik, M. J., & Dekoeijer, G. (2006). Ship transport of CO_2 technical solutions and analysis of costs, energy utilization, exergy efficiency and CO_2 emissions. *Chemical Engineering Research & Design, 84*(9), 847–855.

Baciocchi, R., Costa G., & Polettini, A., Pomi, R. (2010). Accelerated mineral carbonation of steel slags. In *Proceedings of the first International Slag Valorisation Symposium*, Leuven, Belgium, April 6–7, 2009 (pp. 51–63). Leuven, Belgium: Katholieke Universiteit Leuven.

Bai, B., Xiaochun, L. I., Liu, Y., & Zhang, Y. (2006). Preliminary study on CO_2 industrial point sources and their distribution in china. *Chinese Journal of Rock Mechanics & Engineering, 25*(2), 2918–2923.

Baoshan Iron and Steel. (2008). *Sustainability report 2007* (116 pp). Shanghai, China: Baoshan Iron and Steel Co. www.baosteel.com

Benson, S. M., & Cole, D. R. (2008). CO_2 sequestration in deep sedimentary formations. *Elements, 4*(5), 325–331.

Bi, X. R., Zhu, R., & Liu, R. Z. (2012). Fundamental study on stainless steel making through CO_2–O_2 mixture blowing. *Steelmaking, 28*(2), 67–70. (in Chinese).

Bi, X. R., Zhu, R., & Lv, M. (2010). Research on mechanism of dust generation in converter. *Industrial Heating, 06*, 13–16. (in Chinese).

Birat, J.-P. (2010). Carbon dioxide (CO_2) capture and storage technology in the iron and steel industry. In *Developments and innovation in carbon dioxide (CO_2) capture and storage technology Vol. 1: carbon dioxide (CO_2) capture, transport and industrial applications* (pp. 492–521). Woodhead Publishing Ltd.

Bock, B., Rhudy, R., Herzog, H., Klett, M., Davison, J., & Ugarte, D. G. D. L. T., et al. (2003). *Economic evaluation of CO_2storage and sink enhancement options*. Office of Scientific & Technical Information Technical Reports.

Boot-Handford, M. E., Abanades, J. C., Anthony, E. J., Blunt, M. J., Brandani, S., Mac Dowell, N., Fernández, J. R., Ferrari, M.-C., Gross, R., Hallett, J. P., et al. (2014). Carbon capture and storage update. *Energy & Environmental Science, 7*(1), 130–189.

Bossie-Codreanu, D., Le-Gallo, Y., Duquerroix, J., Doerler, N., & Le Thiez, P. (2003). CO_2 sequestration in depleted oil reservoirs. In *Paper presented at the Greenhouse Gas Control Technologies—6th International Conference.*

Bruce, E. T, Weisang, F., Allibert, M., & Fruehan, R. (1987a). In *Electric Furnace Conference Proceedings* (p. 293). Chicago: ISS-AIME.

Bruce, E. T, Weisang, F., Allibert, M., Fruehan, R. (1987b). Effects of CO_2 stirring in a ladle. In *45th Electric Furnace Proceedings* (pp. 293–297), December 8–11, 1987, Chicago, USA.

Bui, M., Adjiman, C. S., Bardow, A., Anthony, E. J., Boston, A., Brown, S., Fennell, P. S., Fuss, S., Galindo, A., Hackett, L. A., et al. (2018). Carbon capture and storage (CCS): The way forward. *Energy & Environmental Science, 11*(5), 1062–1176.

Carpenter, A. (2012). *CO_2abatement in the iron and steel industry* (pp. 83–94). IEA Clean Coal Centre.

Chen, J.B., Yang, Z.L, & He D. Z. (2008). *Treatment outside the furnace.* Metallurgical Industry Press (in Chinese).

Dashti, H., Yew, L. Z., & Lou, X. (2015). Recent advances in gas hydrate-based CO_2 capture. *Journal of Natural Gas Science and Engineering, 23*, 195–207.

Davidson, R. M. (2007). Post-combustion carbon capture from coal fired plant-solvent scrubbing. *CCC/125* (58 pp), London, UK: IEA Clean Coal Centre

Dong, K., Zhu, R., & Liu, R. Z. (2014). Influence of bottom blowing CO_2 in LF on molten steel quality and erosion of ventilated bricks. *Journal of University of Science and Technology Beijing, 36*(s1), 226–229. (in Chinese).

Dong, X. D. (2008). Experimental study on converter slag quickly absorbing CO_2. *Steelmaking., 24*, 29–32. (in Chinese).

Doucet, F. J. (2010). Effective CO_2-specific sequestration capacity of steel slags and variability in their leaching behaviour in view of industrial mineral carbonation. *Minerals Engineering, 23*(3), 262–269.

Duc, N. H., Chauvy, F., & Herri, J.-M. (2007). CO_2 capture by hydrate crystallization-a potential solution for gas emission of steelmaking industry. *Energy Conversion and Management, 48*(4), 1313–1322.

Eloneva, S., Teir, S., Salminen, J., Fogelholm, C.-J., & Zevenhoven, R. (2008). Fixation of CO_2 by carbonating calcium derived from blast furnace slag. *Energy, 33*(9), 1461–1467.

Faruque Hasan, M. M., First, E. L., Boukouvala, F., & Floudas, C. A. (2014). A novel framework for carbon capture, utilization, and sequestration, CCUS. In *Proceedings of the 8th International Conference on Foundations of Computer-Aided Process Design* (pp. 98–107).

Favre, E. (2011). Membrane processes and postcombustion carbon dioxide capture: Challenges and prospects. *Chemical Engineering Journal, 171*(3), 782–793.

Figueroa, J. D., Fout, T., Plasynski, S., McIlvried, H., & Srivastava, R. D. (2008). Advances in CO_2 capture technology—The U.S. department of energy's carbon sequestration program. *International Journal of Greenhouse Gas Control, 2*(1), 9–20.

Fogelholm, C. J., Tveit, T. M., Eloneva, S., Said, A., Revitzer, H., Zevenhoven, R., ... & Fält, M. (2009). Reduction of CO_2 emissions and utilization of slag products by producing calcium-based products from carbonate-free slag products (Slag2PCC Plus) (pp. 314–323). Helsinki, Finland.

Fu ZX, Liu SJ, Fu X. 2010. A method of carbon dioxide injec- tion to Blast Furnace. Chinese Patent, CN101871026 (in Chinese).

Gale, J. (2003). Geological storage of CO_2: What's known, where are the gaps and what more needs to be done. In *aper presented at the Greenhouse Gas Control Technologies—6th International Conference.*

Gao, L., Fang, M., Li, H., & Hetland, J. (2011). Cost analysis of CO_2, transportation: Case study in china. *Energy Procedia, 4*(22), 5974–5981.

GCCSI (Global CCS Institue). (2017a). Yanchang integrated carbon capture and storage demonstration project.

GCCSI (Global CCS Institue). (2017b), Large scale CCS projects

Giorno, L., Drioli, E., & Strathmann, H. (2015). The principle of gas separation. In *Encyclopedia of membranes* (pp. 1–6).

Gu, Y. L., Wang, H. J., Zhu, R., Wang, J., Lv, M., & Wang, H. (2014). Study on experiment and mechanism of bottom blowing CO_2 during the LF refining process. *Steel Research International, 85*(4), 589–598.

Gulyaev, M. P, & Filippov, V. V. (2001). The first systems of bottom blowing with inert-gas in electric arc furnace. *Proceedings of the 6-th Congress of Steelmakers* (p. 308).

Gunter, W. D., Perkins, E. H., & McCann, T. J. (1993). Aquifer disposal of CO_2-rich gases: reaction design for added capacity. *Energy Conversion and Management, 34*(9–11), 941–948.

Guo, M. X., & Chen, X. W. (1993). Action mechanism of bottom- blown CO_2 in the bath of combined-blown converter. *Journal of Iron and Steel Research, 1*, 9–14. (in Chinese).

Guo, X. Q, Yuan, L., Zhang, J., Wu, S. Q. (2010). *A method of CO_2utilization as the transmission medium of pulverized coal injection into BF* (Chinese Patent, CN 102382915) (in Chinese).

Hao, G.-P., Li, W.-C., & Lu, A.-H. (2011). Novel porous solids for carbon dioxide capture. *Journal of Materials Chemistry, 21*(18).

Hara, R. D. O., Spence, A. G. R., & Eissenwasser, J. D. (1986). Carbon dioxide shrouding and purging at IPSCO Melt shop. *Iron Steelmaker, 3*, 24–28.

He, P., Zhang, D. M., & Deng, K. W. (1992). Gas stirring technology with bottom blowing in EAF. *Iron Steel, 9*, 65–70. (in Chinese).

Hong, Y. G. (2012). *Experimental study on desilication of hot metal with carbon dioxide injection* (Master thesis). Northeastern University (in Chinese).

IEA. (2016). *20 years of carbon capture and storage: Accelerating future deployment*. France: International Energy Agency.

Inagaki, M., & Kang, F. (2014). Engineering and applications of carbon materials. In *Materials science and engineering of carbon: Fundamentals* (pp. 219–525).

Ismail, A. F., Kusworo, T. D., Mustafa, A., & Hasbullah, H. (2005). Understanding the solution-diffusion mechanism in gas separation membrane for engineering students. In *Paper presented at the Regional Conference on Engineering Education RCEE*.

Jahangiri, H. R., & Zhang, D. (2012). Ensemble based co-optimization of carbon dioxide sequestration and enhanced oil recovery. *International Journal of Greenhouse Gas Control, 8*(5), 22–33.

Jin, G. L. (1985). Application study of LD converter bottom blowing N_2–CO_2. *Steelmaking., 3*, 96–100. (in Chinese).

Jin, R. J., Zhu, R., Feng, L. X., & Yin, Z. J. (2007). Experimental study of steelmaking with CO_2 and O_2 mixed blowing. *Journal of University of Science and Technology Beijing, 29*(S1), 77–80. (in Chinese).

Kim, H., Lee, J., Lee, S., Lee, I.-B., Park, J.-H., & Han, J. (2015). Economic process design for separation of CO_2 from the off-gas in ironmaking and steelmaking plants. *Energy, 88*, 756–764.

Kovscek, A. R. (2002). Screening criteria for CO_2 storage in oil reservoirs. *Petroleum Science and Technology, 20*(7–8), 841–866.

Kunzler, C., Alves, N., Pereira, E., Nienczewski, J., Ligabue, R., Einloft, S., & Dullius, J. (2011). CO_2 storage with indirect carbonation using industrial waste. *Energy Procedia, 4*, 1010–1017.

Larsen, J. W. (2004). The effects of dissolved CO_2 on coal structure and properties. *International Journal of Coal Geology, 57*(1), 63–70.

Lee, S.-Y., & Park, S.-J. (2015). A review on solid adsorbents for carbon dioxide capture. *Journal of Industrial and Engineering Chemistry, 23*, 1–11.

Leung, D. Y., Caramanna, G., & Maroto-Valer, M. M. (2014). An overview of current status of carbon dioxide capture and storage technologies. *Renewable and Sustainable Energy Reviews, 39*, 426–443.

Li, C. B., & Han, Y. (1996). The application of the combined blowing technique with CO_2 bottom blowing to the converters of Anshan Iron and Steel Company. *Steelmaking., 4*, 19–25. (in Chinese).

Li, Q., Liu, R. Z., & Zhu, R. (2015a). Research on decarburization and chromium conservation by blowing carbon dioxide. *Industrial Heating, 44*(4), 24–26. (in Chinese).

Li, Q., Wang, H., & Zhu, R. (2015b). Experimental study on CO_2 as shield gas of continuous casting. *Continuous Casting, 40*(2), 5–9. (in Chinese).

Li, X., Liu, Y., Bai, B., & Fang, Z. (2006). Ranking and screening of CO_2 saline aquifer storage zones in china. *Chinese Journal of Rock Mechanics & Engineering, 25*(5), 963–968.

Liu, L. S. (1983). Study on CO_2 being protection media for side- blown oxygen lance in converter. *Hebei Metall., 3*, 80–81. (in Chinese).

Liu, H., & Gallagher, K. S. (2011). Preparing to ramp up large-scale CCS demonstrations: An engineering-economic assessment of CO_2, pipeline transportation in china. *International Journal of Greenhouse Gas Control, 5*(4), 798–804.

Liu, Y. F., Li, X. C., & Bai, B. (2005). Preliminary estimation of CO_2 storage capacity of coalbeds in china. *Chinese Journal of Rock Mechanics & Engineering, 24*(16), 2947–2952.

Lv, M., Zhu, R., Bi, X. R., & Lin, T. C. (2011a). Research on CO_2 application in BOF steelmaking process. *Journal of University of Science and Technology Beijing, S1*, 126–130. (in Chinese).

Lv, M., Zhu, R., & Bi, X. R. (2011b). Fundamental research on dephosphorization of BOF by COMI steelmaking process. *Iron Steel, 8*, 31–35. (in Chinese).

Mccollum, D. L., & Ogden, J. M. (2006). Techno-economic models for carbon dioxide compression, transport, and storage & correlations for estimating carbon dioxide density and viscosity. In *Working Paper Series*. Institute of Transportation Studies.

McCoy, S. T., & Rubin, E. S. (2008). An engineering-economic model of pipeline transport of CO_2 with application to carbon capture and storage. *International Journal of Greenhouse Gas Control, 2*(2), 219–229.

Merkel, T. C., Lin, H., Wei, X., & Baker, R. (2010). Power plant post-combustion carbon dioxide capture: An opportunity for membranes. *Journal of Membrane Science, 359*(1–2), 126–139.

Metz, B., (2005). "IPCC Special Report on Carbon Dioxide Capture and Storage." Economics & Politics of Climate Change.

Metz, B., Davidson, O., & De Coninck, H. (Eds.). (2005). Carbon dioxide capture and storage: special report of the intergovernmental panel on climate change. Cambridge University Press.

Mohitpour, M., Golshan, H., Murray, A., (2003). In: *Pipeline design & construction* (1st ed.). ASME Press.

Mondal, M. K., Balsora, H. K., & Varshney, P. (2012). Progress and trends in CO_2 capture/separation technologies: A review. *Energy, 46*(1), 431–441.

Ning, X. J., Yin, Z. J., Yi, C., Zhu, R., & Dong, K. (2009). Experimental research on dust reduction in steelmaking by CO_2. *Steelmaking, 5*, 32–34. (in Chinese).

Normanton, A. S. (Steel Times), Wang, X. C. (Translate). (1990). Application of CO_2 in steelmaking process in English steel industry (Vol. 2, pp. 21–27). Wisco Technologies (in Chinese).

Oelkers, E. H., Gislason, S. R., & Matter, J. (2008). Mineral carbonation of CO_2. *Elements, 4*(5), 333–337.

Olajire, A. A. (2010). CO_2 capture and separation technologies for end-of-pipe applications—A review. *Energy, 35*(6), 2610–2628.

Paules, J. R, Grieshaber, K. W, & Martines, M. L. (1987). Ladle stirring and stream shrouding with carbon diodide at Armeco's Midwestern Steel Division. In *70th Steelmaking Conference Proceedings*, Mar 29–Apr 1, 1987 (pp. 129–132), Pittsburgh, USA.

Paules, J. R. (1987). Ladle stirring and stream shrouding with CO_2 at Armco's Midwestern Steel Division. In *The 70th Steelmaking Conference Proceedings* Mar 29–Apr 1 1987, Pittsburgh, America.

Rawlins, C. H., Richards, V. L., Peaslee, K. D., Lekakh, S. N. (2006). Steelmaking slag as a permanent sequestration sink for carbon dioxide. *Steel Times International, 30*(7), 25–26, 28.

Rice, D. D., Law, B. E., & Clayton, J. L. (1993). Coalbed gas—An undeveloped resource. *US Geological Survey Professional Paper, 1570*, 389–404.

Richards, V. L., Peaslee, K., Smith, J. D. (2008). Geological sequestration of CO_2 by hydrous carbonate formation with reclaimed slag. *TRP 9955* (307 pp). Oak Ridge, TN, USA: Office of Scientific and Technical Information. www.osti.gov/bridge.

Seoane, B., Coronas, J., Gascon, I., Etxeberria Benavides, M., Karvan, O., Caro, J., Kapteijn, F., & Gascon, J. (2015). Metal-organic framework based mixed matrix membranes: A solution for highly efficient CO_2 capture? *Chemical Society Reviews, 44*(8), 2421–2454.

Song, J., & Zhang, D. (2013). Comprehensive review of caprock-sealing mechanisms for geologic carbon sequestration. *Environmental Science and Technology, 47*, 9–22.

Spigarelli, B. P., & Kawatra, S. K. (2013). Opportunities and challenges in carbon dioxide capture. *Journal of CO_2 Utilization, 1*, 69–87.

Suzuki, K., Kumagai, T., & Sano, N. (1992). Removal of boron from metallurgical-grade silicon by applying the plasma treatment. *ISIJ International, 32*(5), 630–634.

Svensson, R., Odenberger, M., Johnsson, F., & Strömberg, L. (2004). Transportation system for CO_2 – application to carbon capture and storage.*Energy Conversion and Management, 45*(15–16), 2343–2353.

Tang, F. P, Zhao, G., Li, D. G., Cao, D., & Zhao, C. L. (2011). *A smelting method for stainless steel production with CO_2blowing in AOD* (Chinese patent, CN102146499) (in Chinese).

Teng, Y., & Zhang, D. (2018). Long-term viability of carbon sequestration in deep-sea sediments. *Science Advances, 4*.

Teir, S., Eloneva, S., Fogelholm, C.-J., & Zevenhoven, R. (2007). Dissolution of steelmaking slags in acetic acid for precipitated calcium carbonate production. *Energy, 32*(4), 528–539.

Uibu, M., Kuusik, R., Andreas, L., & Kirsimäe, K. (2011). The CO_2-binding by Ca–Mg–silicates in direct aqueous carbonation of oil shale ash and steel slag. *Energy Procedia, 4*, 925–932.

Van Wijngaarden, M. J., & Pieterse, A. T. (1994). Bottom-stirring in an electric-arc furnace: Performance results at Iscor Vereeniging Works. *Journal of the Southern African Institute of Mining and Metallurgy, 1*, 27–34.

Vandeginste, V., & Piessens, K. (2008). Pipeline design for a least-cost router application for CO_2, transport in the CO_2, sequestration cycle. *International Journal of Greenhouse Gas Control*, *2*(4), 571–581.

Wang, H., Zhu, R., & Liu, R. Z. (2014). CO_2 application on bottom blowing in EAF. *Industrial Heating, 43*(2), 12–14. (in Chinese).

Wang, H. J., Teng, L. D., Seetharaman, S. (2012). Heat and material balance calculations with introduction of CO_2 during EAF processes. In: 5th International Congress on Steelmaking Technology, October 1–3, 2012, Dresden

Wang, H. J., Viswanathan, N. N., Ballal, N. B., & Seetharaman, S. (2009). Modelling of reactions between gas bubble and molten metal bath-experimental validation in the case of decar-burization of Fe–Cr–C melts. *High Temp Mater Process, 28*(6), 407–419.

Wang, H. J., Viswanathan, N. N., Ballal, N. B., & Seetharaman, S. (2010). Modelling of physico-chemical phenomena between gas inside a bubble and liquid metal. *Journal of Chemical Reaction Engineering, 8*, 45–54.

Wang, S. L. (1986). Metallurgical behaviours of gas bottom blowing in converter. *Steelmaking, 3*, 24–30. (in Chinese).

Wang, S. L. (1989). Gas supply model of bottom blowing CO_2 in converter. *Steelmaking., 4*, 1–7. (in Chinese).

Wilberforce, T., Baroutaji, A., Soudan, B., Al-Alami, A. H., & Olabi, A. G. (2019). Outlook of carbon capture technology and challenges. *Science of the Total Environment, 657*, 56–72.

Wilcox, J., Haghpanah, R., Rupp, E. C., He, J., & Lee, K. (2014). Advancing adsorption and membrane separation processes for the gigaton carbon capture challenge. *Annual Review of Chemical and Biomolecular Engineering, 5*, 479–505.

Wu, J. J., Li, Y., Ma, W., Liu, K., Wei, K., Xie, K., Yang, B., & Dai, Y. (2014). Impurities removal from metallurgical grade silicon using gas blowing refining techniques. *SILICON, 6*(1), 79–85.

Wu, J.-J., Ma, W.-H., Yang, B., Dai, Y.-N., & Morita, K. (2009). Boron removal from metall-urgical grade silicon by oxidizing refining. *Transactions of Nonferrous Metals Society of China, 19*(2), 463–467.

Xiuzhang, W. (2014). Shenhua Group's carbon capture and storage (CCS) demonstration. *Mining Report, 150*(1–2), 81–84.

Yi, C., Zhu, R., Yin, Z. J., & Hou, N. N. (2009). Experimental study of COMI steel making process based on 30t converter. *Chinese Journal of Proceeding Engineering, S1*, 222–225. (in Chinese).

Yin, Z. J., Zhu, R., & Yi, C. (2009). Fundamental research on controlling BOF dust by COMI steelmaking process. *Iron Steel, 10*, 92–94. (in Chinese).

Yu, H. C, Wang, H. J., & Chu, S. J. (2015). Fundamental investigations on M–LCFeCr smelting by introducing CO_2. In *The 6th International Congress on the Science and Technology of Steelmaking (ICS2015)*, May 12–14, 2015, Beijing.

Yu, H. C. (2015). Master thesis. Beijing: University of Science and Technology (in Chinese).

Yuan, Z. F, Zhao, H. X., Pan, Y. F. (2007). *Method for CO_2being applied in slag splashing in converter* (Chinese patent). ZL200610165219.2 (in Chinese).
Zhang, Z. X., Wang, G. X., Massarotto, P., & Rudolph, V. (2006). Optimization of pipeline transport for CO_2 sequestration. *Energy Conversion and Management, 47*(6), 702–715.
Zhao, H.-X., Yuan, Z.-F., Wang, W.-J., Pan, Y.-F., & Li, S.-Q. (2010). A novel method of recycling CO_2 for slag splashing in converter. *Journal of Iron and Steel Research International, 17*, 11–16.
Zhao, X. J. (1995). Development of bottom blowing technology of EAF abroad. *Steeltubes, 5*, 12–15. (in Chinese).
Zhu, R., Lv, M., Lin, T. C. (2012). Application research of carbon dioxide during the steelmaking process. In *5th International Congress on Steelmaking Technology*, October 1–3, 2012, Dresden.
Zhu, R., Yi, C., Chen, B. Y., & Wang, C. Y. (2010). Research on inner circulation control of steelmaking dust by COMI process. *Energy Metall Industries, 1*, 48–51. (in Chinese).

CCUS: What is It? How Does It Cost? Techno-Economic Analysis

4

Qianguo Lin, Xi Liang, Ming Lei, Y. M. Zhang, Y. R. Pan, and N. Wang

CO_2 capture, utilization and storage technology (CCUS) is one of the few technologies that can help industry achieve large-scale CO_2 emission reductions. Currently, huge investment and operating costs, as well as lack of financially viable business models are generally seen as the main obstacles to the development of CCUS. According to IEA estimates, currently for CCS project to sequester one ton of CO_2 emissions, it would cost as high as \$200–300 on average, and that cost varies greatly, depending on local and project-specific engineering conditions. A full-chain CCUS project is particularly complex, involving multiple interactions between carbon capture, transportation, utilization and storage activities. Even within an iron or steel plant, CCUS involves not only multiple gas sources in the capture process, but also different capture technologies for different gas sources. That being the case, cost estimates in connection with tech-performance of CCUS are all the more important, because they can help decision-makers understand and identify cost-effective pathways, when making decisions about deploying CCUS projects.

The original version of this chapter was revised: A sentence has been added on page 109. The correction to this chapter can be found at https://doi.org/10.1007/978-981-16-4310-1_8.

M. Lei (✉)
Guanghua School of Management, Peking University, Beijing, China
E-Mail: leiming@gsm.pku.edu.cn

Q. Lin · X. Liang
University of Edinburgh Business School, Edinburgh, U.K.

Y. M. Zhang · Y. R. Pan · N. Wang
North-China Electric Power University, Beijing, China

J. Fu et al. (eds.), *Climate Mitigation and Adaptation in China*,
https://doi.org/10.1007/978-981-16-4310-1_4

Thus, building upon the previous chapter, in Chapter 4, we attempt a systematic techno-economic analysis of the efficacy of different carbon capture technologies for major sources of carbon emissions in the steel sector.

Chapter 4 is organized into four sections. Section 1 identifies the main CO_2 emission sources of a typical production process of China's iron/steel industry. Section 2 discusses suitable carbon capture technologies for each emission source. Section 3 conducts a systematic techno-economic analysis of different capture methods for each emission source. Section 4 concludes with a caveat, calling for attention to local and firm-specific engineering conditions.

1 Section 1

1.1 Main CO_2 Emission Sources from Iron/steel Industry

The steel/iron production process can be divided into a main process and an auxiliary process. The main process can be further categorized into long process (BF/BOF) and short process (EAF). The long process (BF/BOF) starts with raw materials such as iron ore, through the processes of mining, milling of ores, blast furnace, converter secondary metallurgy, continuous casting, steel rolling and other processes. The short process (EAF) refers to the use of scrap steel as raw materials to make steel product, though casting of electric furnace, ladle furnace, continuous rolling and other processes. The auxiliary process refers to the production process which has a great influence on the environmental, such as coking, boiler, oxygen production, fuel gas processing, energy discharge and internal transportation. At present, the long process is mainly adopted by China's iron/steel industry; crude steel production accounts for 90% of total steel production in China (Cui, 2016); in addition, the per ton steel CO_2 emission of crude steel produced by a short process is far less than that of crude steel produced by a long process (CO_2 emission associated with converter process is 5 times than that associated with electrical furnace process (Ran & Weng, 2006). Therefore, the CO_2 produced by long process production process is the main CO_2 emission source of China's steel/iron industry.

The production processes of a long process steel/iron industry mainly include 5 processes: coking, sintering, iron-making, steel-making and steel rolling. The iron-making process has three types: blast furnace iron-making, direct-reduction iron-making and molten reduction technology. The steel-making process has two types, which is converter steel-making and electric furnace steel-making. In the process of coking, sintering, iron-making, steel-making and rolling steel, fuel coal, coking coal, coke, natural gas and fuel oil are not only the source of energy, but also raw materials. Some part of carbon in these energy and material sources is converted and released in the form of CO_2 by the way of reduction or oxidation reaction in the whole production process. The main CO_2 emission sources are released from units include lime kiln, coke oven, hot blast furnace, blast furnace, converter and self-provided power plant.

1.2 CO_2 Emission from Lime Kiln

In iron-making process, lime kiln is used to produce lime as flux and slag materials to be used in blast furnace iron-making and converter iron-making processes to remove harmful elements from smelting, such as sulphur, silicon, phosphorus other impurities. Lime is formed by the limestone calcination, the main component of limestone is calcium carbonate. The basic principle of burning limestone can be expressed by the following chemical reaction equation (Eq. 1):

$$CaCO_3 \xrightarrow{>900\ ^\circ C} CaO + CO \uparrow \tag{1}$$

The limestone and air are preheated to about 900 °C in the lime kiln preheater. After lime produced, the lime (product) will be cooled to below about 100 °C by cooling gas. After the cooling gas enters a roast area to mix with the exhaust gas from combustion, the temperature of mixed gas will reach about 900 °C. The mixed gas is then recycled and reused for heating limestone and air in the subsequent limestone preheating process. After the preheating process, the temperature of flue gas is reduced to 60–80 °C. Although when the breakdown of limestone can release about 40% of its weight in the form of CO_2, after mixing with exhaust gas from combustion and circulating, CO_2 content in the flue gas will be significantly decreased. CO_2 from different lime kiln unit may have different concentrations.

According to the report of *Applying carbon capture and storage to a Chinese steel plant* which is prepared for Global Carbon Capture and Storage Institute Ltd by Toshiba International Corp Pty Ltd/Toshiba Corporation in collaboration with Tongfang Environment Co. Ltd, China Shougang Caofeidian plant has 6 sets shaft lime kilns, used for production of lime needed by steel mill. The flue gas composition of #4 lime kiln from Caofeidian plant is given in Table 1, indicating 19.3% CO_2 concentration. Shen (2008) reported that the CO_2 concentration of flue gas

Table 1 Flue gas composition of #4 Lime Kiln in Shougang Jingtang

	Flow rate	Temperature	Concentration (CO_2)	CO_2 emission	
	m^3N/h	°C	Vol.%	t/d	kt/year
#4 lime kiln	75,304	110	19.3	685	205

of a typical limestone kilns outlet in the integrated iron and steel enterprise was 15–21%.

1.3 CO_2 Emission from Coke Oven

The coke oven flue gas from steel/iron plant is the exhaust gas produced by combustion and discharged through a coke oven chimney. The mixture of coke oven gas and blast furnace gas is generally used for coke oven heating by integrated iron and steel enterprises. Accordingly, the CO_2 concentration is mainly affected by the combustion condition and the composition of combustion feed gas. According to the report of *Applying carbon capture and storage to a Chinese steel plant* which is prepared for Global Carbon Capture and Storage Institute Ltd by Toshiba International Corp Pty Ltd/Toshiba Corporation in collaboration with Tongfang Environment Co. Ltd, the CO_2 concentration of the combustion flue gas from coke oven in China Shougang Jingtang was 11%. The report of Global Carbon Capture and Storage presented a concentration range of 10–15% (Global CCS Institute, 2016).

There is another gas source related to coking produced by coking process. It is called raw coke oven gas or raw coal gas. The raw coal gas can be purified though a coke oven gas purification system (Fig. 1). Coke oven gas purification process is differentiated mainly based on different desulphurization and deamination configurations. Generally, there are two kinds desulfurization system: dry desulfurization and wet desulfurization. Wet desulfurization process can be further categorized into two types: wet oxidation process and wet absorption process. The wet absorption process has the function of removing CO_2 simultaneously, leading to a low CO_2 concentration in the purified flue gas.

The Sulfiban method of wet absorption process was applied in the coal gas refining system in Baosteel phase II which was put into operation in May 1997. Through an intermediate cooling tower, the coal gas after naphthalene removed enters an ammonia absorption facility, where a 75% of ammonium dihydrogen phosphate is used as absorption solution. The rich solution is ammonium

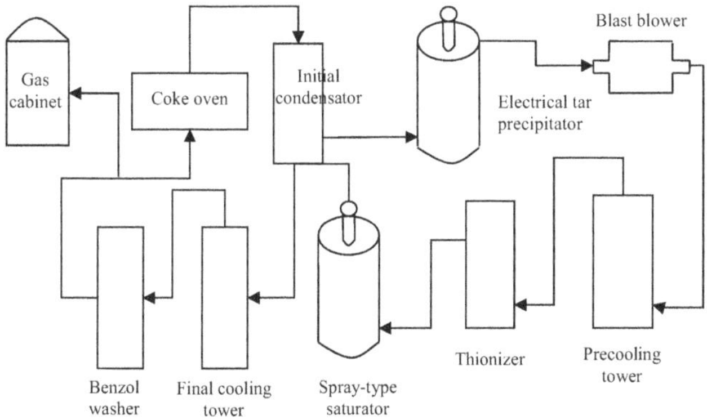

Fig. 1 Coke oven gas purification system

hydrogen phosphate. Ammonia solution with concentration of 18% is obtained after desorption. Anhydrous ammonia with the concentration of 99.8% is finally obtained after further pressurized and concentrated. The absorbents can be recycled after desorption. The coal gas which absorbed ammonia enters a final cooling tower after dephenolization and decyanation. The coal gas without ammonia enters a benzol scrubber to remove crude benzene. The gas is then delivered to a gas cabinet for storage after desulfuration and decyanation processes. The sulfur enriched solution is desorbed in a desorption facility. The obtained H_2S, HCN and CO_2 are then sent to combustion furnace to generate SO_2 and NO_x.

The composition of a typical purified coal gas is illustrated in Table 3. Because CO_2 is partially removed by ammonia absorption process, the concentration of CO_2 is as low as 2.1%. Comparing coke oven flue gas and coke oven gas, the former is a combustion product discharged through a coke oven chimney, its CO_2 concentration is relatively high; the latter is a direct product produced in the coking process of coal, its CO_2 concentration is relatively low particularly after absorption of ammonia. Therefore, coke oven flue gas is more suitable for the application of carbon capture technology than the coke oven gas (Table 2).

Table 2 Composition of typical coke oven gas

Composition	CO	CO_2	H_2	N_2	CH_4	C_2H_4	C_2H_6	C_3H_6	O_2	H_2O
Coke oven gas	5.9	2.1	58.8	3.8	25.6	2.3	0.9	0.2	0.4	saturation

Table 3 Operating condition at TRT inlet and outlet

Item	Numerical value
Blast furnace top pressure	0.20 MPa
TRT inlet pressure	0.17 MPa
TRT outlet pressure	0.01 MPa
TRT inlet gas temperature	100 °C
TRT outlet gas temperature	50 °C

1.4 CO_2 Emission from Blast Furnace

An iron-making process makes use of iron ore, fuel (coke, sometimes it is supplemented by heavy oil, pulverized coal, natural gas etc.) and other auxiliary raw materials to produce molten iron and two by-products (slag and coal gas) through heating, decomposing, revivification slagging, desulfuration and other reactions in a blast furnace. In the production of blast furnace, iron ore, coke and limestone are put into the furnace from the top. After preheated through a hot blast furnace, air is blown into the furnace from the bottom by a blast blower. Under high temperature, carbon of coke, pulverized coal, heavy oil, natural gas and other fuels and blowing oxygen burn to generate carbon monoxide (CO). In the rising process of CO in a furnace, the oxygen in iron ore is removed, the iron ore is reduced to molten iron. The molten iron is released from a iron hole as the product of blast furnace; the non-reductive impurities in iron ore react with the fluxes such as limestone to generate slag, discharging from a slag hole; the generated gas (blast furnace gas) is released from the top of the furnace. After removal of dust, the gas can be used as the fuel of hot blast furnace, heating furnace, coke oven, boiler, etc. The pressure of blast furnace gas (BFG) mainly depends on the pressure of blowing air.

Blast furnace gas can be classified into coarse dust removal coal gas, semi-clean dust removal coal gas and fine dust removal coal gas. According to different purification methods, blast furnace gas can be purified by dry dust and wet dust removal methods. The coarse dust removal is usually conducted with dry method. The coarse dust removal is the first purification in a dust removal process. The semi-clean dust removal can adopt wet method or dry method. The fine dust removal is the final stage of BFG dust removal. The purified fine gas can be sent to a gas pipeline after energy recovered by a blast furnace top gas pressure recovery turbine system (TRT). Blast furnace top gas pressure recovery turbine (TRT)

Table 4 BFG composition of typical iron/steel plant in China

Composition	CO	CO_2	H_2	N_2	O_2
Vol%	21.8	23.1	3.2	51.7	0.2

system makes use of pressure energy and sensible heat of BFG. Through the system, the blast furnace gas pressure energy and part of heat energy can be converted into mechanical energy and then drives a generator to generate electricity.

The capacity of TRT should be matched to the gas amount of a blast furnace and the furnace top pressure, while amount of blast furnace gas is associated with the capacity of blast furnace blower. Temperature and pressure will be mainly changed between the inlet and outlet of TRT. Table 3 illustrated a typical operating condition at TRT inlet and outlet. It is observed that pressure and temperature decrease significantly after TRT. The main composition of blast furnace gas after TRT is CO, CO_2, N_2, H_2 and CH_4. Table 4 is the blast furnace gas composition of a typical steel/iron plant in China. The CO_2 concentration is 23.1% which is higher than that of flue gas from lime kiln.

1.5 CO_2 Emission from Converter

In the Linz-Donawitz smelting process, the molten iron in the furnace reacts with the blown oxygen. The produced gas is called Linz-Donawitz gas (LDG), which is mainly from the oxidation of carbon in molten iron. Due to the high temperature in a converter furnace, the main oxide of carbon is CO, account for about 90% of total amount carbon oxides. A small portion of carbon reacts directly with oxygen to produce CO_2 and some CO could escape from the surface of molten steel and react with oxygen to produce CO_2.

At the beginning of blowing oxygen process, the temperature of molten iron is low, the easily oxidized elements silicon and manganese in the molten iron are first oxidized, then phosphorus, sulfur and other elements; meanwhile, a small portion of iron is also oxidized. Because carbon has a lower oxidation rate, furnace gas volume produced is relatively small and the furnace gas temperature is relatively low. With huge heat released associated with the oxidation reaction of above elements, the temperature of the molten pool increases correspondingly. When the temperature of the molten pool exceeds 1400 °C, carbon and oxygen react violently, leading to CO content in the furnace gas gradually increased. The amount of furnace gas produced per unit time is related with the intensity of blowing oxygen and carbon reduction speed. The greater of the intensity of blowing oxygen and the faster of the carbon reduction speed, the more of the amount of furnace gas produced in unit time. In the later period of converting, with the decrease of carbon in molten iron, the amount of LDG production decreases, but the temperature of LDG rises with the rising temperature of molten pool.

The Linz-Donawitz steel-making is usually intermittent. One process may take about 40 min. In the time of beginning and end of converting, the amount of furnace gas is small and CO content is low, hence, the furnace gas is normally taken in the middle section. The former and latter periods are called pre-combustion period and post-combustion period. The middle period for capture furnace gas is called capture period. Due to the periodicity of a Linz-Donawitz steel-making process, the capture time is intermittent, resulting in furnace gas intermittently discharged.

In the process of converting, when adding auxiliary raw material to a converter, the amount of furnace gas will increase rapidly. The auxiliary raw materials are slagging flux and coolant. The slagging flux mainly uses lime and the coolant mainly uses steel scrap, iron ore and oxidized scale. After the gas leaves the furnace mouth of converter, it will burn when meets air. In addition, the gas contains highly toxic materials. Therefore, the high temperature flue gas needs to be captured at the furnace mouth of converter. When the high temperature flue gas pass through a evaporation cooling facility, the flue gas is initially cooled while some heat energy is recovered. The temperature at the outlet the evaporation cooling facility is about 900 °C (Zhang et al., 2017). After initially cooling and coarse dust removal, the flue gas needs to be further cooled and depressurization to meet operating conditions of fine dust removal units.

Dust removal purification methods have two types: wet method and dry method. ① The typical process of wet purification system is: after released from a converter, the gas is initially cooled to 800–1000 °C through a vaporizing cooler, then go through the first-degree venturi tube, the first elbow dehydrator, the second-grade venturi tube and the second elbow dehydrator successively. Spraying water at the throat of the venturi tube will decrease gas temperature to about 35 °C and reduce the dust content in the gas. The purified gas is then sent to a gas storage cabinet by an induced draft fan. ② A typical process of dry purification system is: the gas temperature is dropped to 1000 °C through vaporizing cooler, then further dropped to 200 °C by using an evaporative cooling tower. Through dust removal of dry-type electrostatic precipitator, the purified gas is fed

Table 5 Composition of typical LDG

Composition	CO	CO_2	H_2	N_2	CH_4	C_2H_4	C_2H_6	C_3H_6	O_2	H_2O
Vol%	58.5	15.1	3.7	21.5	1.0	0.1	0	0	0.1	saturation

into a gas storage cabinet by an induced draft fan. Compared to wet method, dry method purification system has the following advantages:

1. The dust content of gas is low;
2. Because of well-developed automation system leading to a quick switch of gas capture system, gas capture rate is increased. Normally, the capture rate of dry method system is 15–30% more than that of wet method one.
3. The dust content of gas is low after purification, resulting in longer life of draught fan and less workload of maintenance s;
4. System resistance is small;
5. Water conservation;
6. Total power load of dry method system is much less than wet method;
7. Increase the recycled amount of dust containing iron and reduce the associated transportation cost;
8. Water treatment system is unnecessary, reduce land use;
9. Reduce secondary pollution of waste water and reduce dispersed gas amount.

The disadvantage is:

1. Complex system, high failure rate and high maintenance cost;
2. Requires high management and operational skills. More training for manager and operators be carried out.
3. Fixed investment cost is high (Wang et al., 2008).

Table 5 demonstrates the composition of a typical Linz-Donawitz gas (LDG) after dust removal. The temperature of the gas is about 100 °C and the pressure is slightly higher than the ambient pressure.

1.6 CO_2 Emission from Hot Blast Furnace

An iron-making blast furnace is usually equipped with several hot blast furnaces, with each hot blast furnace consisting of a combustion chamber, a regenerative chamber, a wind mixing chamber, and a variety of valves and corresponding pipes. The air required for the combustion of a hot blast furnace is provided by a combustion fan. According to the regulations of a hot blast furnace operating instruction, in the normal operating process of a hot blast furnace, a smothered furnace state is necessary between blowing-in and combustion operations. There is no direct transfer between the blowing-in and the combustion state. In short, the production cycle of a hot blast furnace should be combustion—stop—blowing-in—stop—combustion (Liu, 2016).

The combustion stage is also called a burning furnace stage. The aim is to heat hot blast furnace lattice bricks to a certain temperature to reserve heat by lattice bricks. The combustion fuel is mainly of blast furnace gas, mixture of blast furnace gas and coke oven, gas or mixture of blast furnace gas and CFG as well as combustion air. Those fuels and air are fed into hot furnace through the burner with a certain ratio between the fuels and the air. The fuels burn in the burner aided with air. The resulting gas is discharged from a flue outlet. At combustion stage, cold air inlet and hot air outlet are both in a closure state. After the combustion process (i.e., lattice bricks storage heat sufficiently), and the operation moved to the next air blowing-in stage.

At the stage of blowing-in, cold air is heated when fed into the blast furnace. The burner and the flue outlet are kept off, while cold air inlet and hot blast air outlet are open. With the heat exchange between cold air and lattice bricks, the temperature of the lattice bricks decreases. Finally, the operation is ceased, waiting for the combustion in the next round.

Because the ratio of air and fuel in the hot blast furnace is different and the inlet pressure and temperature of gas and air are also different, the CO_2 content would be different in the flue gas of hot blast furnace. The residual O_2 is also affected by air–fuel ratio. The general composition of hot blast flue gas is shown in Table 6. Although the hot blast flue gas produced by a single hot blast furnace is continuous, in the actual production process, the emission of flue gas at outlet of a chimney is approximately continuous as multiple hot blast furnaces share a flue gas discharge chimney.

Table 6 Composition of typical hot blast furnace flue gas

Composition	CO_2	N_2	O_2	H_2O	Dust /(mg m^{-3})
Vol%	28.52	65.15	2.51	3.82	5

1.7 CO_2 Emission from Power Generation Unit

The power plant in a steel/iron enterprise is to supply heat and electricity to meet iron/steel making demand. The power plant normally consumes the combustible gases produced during iron/steel making processes, including coke oven gas and blast furnace gas, to improve power supply reliability, to reduce power generation cost, and to contribute to energy conservation and environmental protection of iron/steel industry. The CO_2 concentration of flue gas produced by power and heat generation processes in the power plant is a little low, being about 10–15%.

1.8 Summary of main CO_2 sources in iron/steel industry

There are multiple CO_2 emission sources with huge emissions in an integrated steel/iron plant. The different emission sources correspond to different production processes, thus have different emission characteristics. Among them, the CO_2 content of flue gas of lime kiln, hot blast furnace gas, blast furnace gas and converter furnace gas is higher than that of post-combustion flue gas generated from a typical coal-fired power plant; while, the CO_2 concentration of coke oven flue gas and the self-provided power plant could be lower than that of post-combustion flue gas from coal-fired power plant.

In terms of pressure, except that there is a higher pressure in blast furnace gas before TRT, the pressure of other emission sources is close to an atmospheric pressure. Regarding temperature, the flue gas from hot blast furnace normally has high temperature than other emission sources, thus heat recovery is desired before CO_2 capture. As for discharge way, the discharge of LDG is intermittent, while others are continuously or approximately continuously discharged. Among all gas sources, LDG and BFG are both combustible. Their calorific value can be increased accordingly by capturing CO_2. In general, the different emission characteristics of CO_2 enriched gases from iron/steel industry have influence on the technical and economic performance of different capture technologies.

2 Section 2

2.1 Carbon Capture Technologies for Steel/iron Industry

The technology development in China's steel/iron industry is unbalanced, although the overall energy consumption is much higher than that of the international community. With the implementation of energy conservation technologies in China's steel/iron industry, the energy consumption level of China's large steel/iron companies is already close to that of international iron/steel enterprises. Therefore, the indirect emission reduction through energy conservation is more and more limited for those large enterprises, resulting in the emission reduction cost through energy conservation is getting higher and higher. It is economically hard for those large steel/iron companies to meet their long-term emission reduction target. As a large-scale CO_2 emission reduction technology with great potential, carbon capture technology can be a cost-effective option to the indirect emission reduction technology. It can benefit large-scale enterprise by achieving a sustainable emission reduction target without changing its already established energy balance system. Therefore, carbon capture technology is increasingly recognized by large steel/iron enterprises in the world.

According to different CO_2 sources, carbon capture technologies are generally categorized into three types: post-combustion capture, pre-combustion capture and oxyfuel combustion capture. Among them, the post-combustion capture refers to the capture of CO_2 from the flue gas of fossil fuel combustion. According to the analysis in Sect. 1, the most of CO_2 associated with steel/iron production is mainly from post-combustion flue gas, hence, this study focuses on the post-combustion capture technology for iron/steel industry. Based on different separation principles, the post-combustion capture can be further typed into absorption capture technology, adsorption capture technology, membrane capture technology, cryogenic capture technology, hydrate capture technology and mineral carbonation capture technology.

2.2 Absorption Capture Technology

Absorption capture or separation technology is a technology that uses absorbent solution to separate CO_2 from mixed gases containing CO_2. It can be divided into physical absorption method and chemical absorption method (Li, 2014). The absorption technology has been widely used in chemical industry, oil refining,

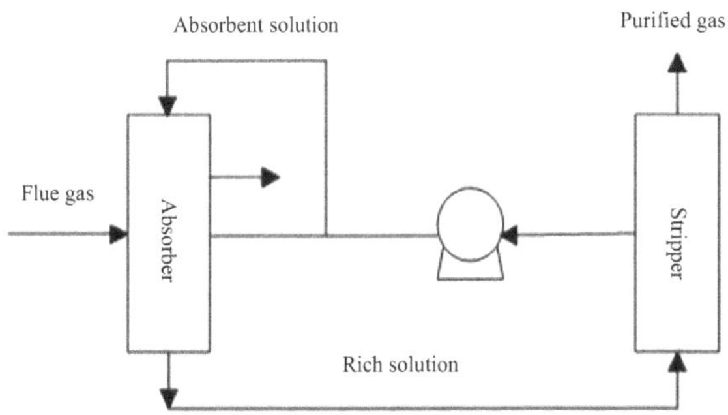

Fig. 2 Principle of chemical absorption method

natural gas processing and other fields, and it is proved that it has the application prospect in steel/iron industry (IEA, 2010) (Fig. 2).

The chemical absorption method refers to the use of alkaline chemical absorbent, such as ammonia, to absorb CO_2 in a flue gas by chemical reaction and form unstable salt. Under a certain temperature and pressure condition, the salt can be decomposed in reverse to release CO_2, so as to achieve the aim that removing CO_2 from the flue gas (Jiang, 2015). At present, the typical chemical absorbents include mono ethanol amine (MEA), alkyl alcohol amine, ammonium hydroxide, hot potassium carbonate solution, etc. (Sun, 2009), the recovery rate of CO_2 can reach 90%, the purity can reach above 99%.

The study of Gielen (2003) indicated that chemical absorption methods can be used to deal with blast furnace gas (BFG), Linz-Donawitz gas (LDG), and flue gas produced in the process of direct reduction and melting. Tseitlin et al. (1994) used a mono ethanol amine (MEA) solvent to separate CO_2 from blast furnace gas. Tobiesen et al (2007) adopted different amine solvents to separate CO_2 from blast furnace gas. In order to separate the CO_2 with the concentration of 30% from hot blast furnace, Cheng et al (2010a) investigated the performance of different solvents. Yoon et al (2011) developed a potassium carbonate/cyclic diamine solvent to separate CO_2 from the flue gas with concentration of 15–20% in the steel/iron industry.

In technology demonstration field, POSCO developed an ammonia absorption technology to separate CO_2 form blast furnace gas. In order to reduce the cost of CO_2 separation, a medium and low temperature waste heat from steel/iron mill was used as the desorption heat of CO_2 (Kim et al., 2009). The first phase of a pilot project was completed in 2008 with processing capacity being 50m^3/h,

CO_2 recovery rate being above 90% and CO_2 volume fraction (concentration) being above 95%. Base on phase one, the second phase of the pilot project was completed in 2011 with blast furnace gas processing capacity being expanded to 1000 m^3/h, CO_2 recovery rate being over 90%, and the concentration of CO_2 exceeding 99.5% (POSCO, 2013). In the project of COURSE50, in order to reach the target of 30% emission reduction, Japan New Energy Industrial Technology Development Organization (NEDO) commissioned the Nippon Steel & Sumitomo Metal and JEF to carry out relevant technology development under the support of the Japan Iron and Steel Federation (Li and Lu, 2009).

Nippon Steel & Sumitomo Metal developed ESCAP (energy conversation CO_2 absorption process) low energy consumption CO_2 separation process, which uses the developed new type of amine solvent (RN) to remove CO_2 from blast furnace gas, the desorption heat is low temperature waste heat that has not been utilized in previous iron works (TNCSTEEL, 2015). The process required a low regeneration temperature, and the newly developed amine solvent is less corrosive to equipment and devices. For the combined use of the new type amine solvent RN and the waste heat recovery system, CO_2 capture cost was expected to be almost 50% lower than that of MEA method (30%MEA) (Goto et al., 2013). Toshiba International Corp Pty Ltd/Toshiba Corporation in collaboration with Tongfang Environment Co. Ltd proposed an amine solvent absorption method to capture CO_2 in the case of Shougang Jingtang steel plant in China.

The physical absorption method is the absorption of CO_2 by means of physical dissolution. Its advantages include good absorption effect, low energy consumption and high separation recovery rate. Its disadvantages is that it is only suitable for flue gas with high CO_2 content (Huang, 2013). Therefore, for low CO_2 content emission sources of steel/iron plant, the physical absorption method may not be technically feasible. The technologies developed by typical physical absorption method include Selexol™ system based on ethylene glycol and Rectisol® system based on methyl alcohol. The Selexol process can operate at an ambient temperature, while the Rectisol process operates as low as -60 °C temperature condition. These systems have been applied in a large scale in the process of syngas purification and natural gas processing.

2.3 Adsorption Capture Technology

Adsorption capture technology is a technology that utilizes the selectivity of the solid adsorbent for CO_2 and reversible adsorption to separate CO_2 from mixed feed gases containing CO_2. The adsorbent adsorbs CO_2 in a low temperature (or high pressure). The CO_2 is de-adsorbed when temperature increasing (temperature swing adsorption, TSA), pressure reduction (pressure swing adsorption, PSA or vacuum pressure swing adsorption, VPSA) or imposing low voltage current (electricity swing adsorption, ESA). Through periodic changes of temperature (or pressure), the CO_2 is separated from a gas stream (Han et al., 2009). Currently, pressure swing adsorption process (PSA and VPSA) is more widely applied than TSA (Kuramochi et al., 2011).

The basic principle of pressure swing adsorption (PSA) is that using the difference of adsorption properties of a kind of gas by a solid material and the characteristic that the adsorption capacity varies with pressure. Through a periodic pressure transformation process, the separation or capture of a kind of gas can be realized. The adsorbate adsorb a specific gas (such as CO_2) at a higher pressure, then the desorbed by reducing pressure. The process of adsorption at high pressure and desorption at normal pressure is normally called pressure swing adsorption (PSA); the process of adsorption at normal pressure or slightly higher than normal pressure, and desorption through vacuum is called vacuum pressure swing adsorption (VPSA).

The performance of adsorbent determines the process and cost of pressure swing adsorption method. The pressure swing adsorption can use carbon-based adsorbent, zeolite and hydrotalcite-like compounds as the adsorbent and is widely applied in purification of hydrogen and CO_2. It can also be used to produce food grade CO_2, purification of CO, purification of natural gas, and oxygen and nitrogen separation.

In the project of COURSE50, JFE steel company developed a two-stage pressure swing adsorption method (zeolite adsorbents) to separate CO_2 from blast furnace gas (Qi et al., 2013). The pilot system (ASCOA-3) was developed with CO_2 processing capacity of 3t/d. The system separates the flue gas with CO_2 concentration of 33%, the recovery rate of CO_2 reaching 80% and the CO_2 purity being 99%. In 2011, POSCO developed a pressure swing adsorption technology (PSA) technology to separate the mixture of CO and CO_2. A small pilot with processing capacity of $1Nm^3/h$ and the concentration of captured CO_2 being over 99% was established (POSCO, 2013).

2.4 Membrane Capture Technology

Membrane capture technology can be typed into membrane separation method and membrane-absorption method (Guo et al., 2016). Membrane separation method separates mixed gases according to the difference in permeation selectivity of

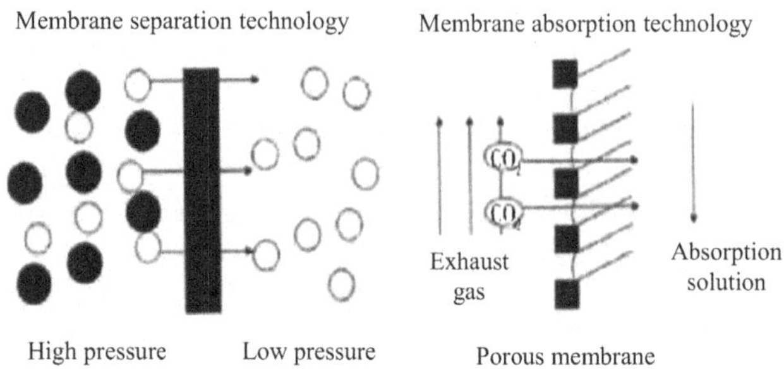

Fig. 3 Schematic principle of membrane separation CO_2

membrane materials for different gases. The driving force of membrane separation is the pressure difference on both sides of the membrane. The gas with high permeability preferentially passes through the membrane and form a permeating airflow. The gas with low permeability forms a residual airflow on the other side. Two streams are led respectively to achieve gases separation (Feron & Jansen, 2002). The ability of this method to separate CO_2 mainly depends on the selectivity of membrane material for CO_2 (Guo et al., 2016). The membrane materials for separation of CO_2 are high polymer material, inorganic material, hybrid organic–inorganic materials etc. (Wang et al., 2016). The separation process can adopt a single-stage membrane separation or a multistage membrane separation process (Fig. 3).

The membrane absorption method is driven by the pressure difference on both sides of membrane, depending on adsorption capacity of each component of gas on the membrane surface and the differences in dissolution and diffusion in membrane (difference of permeation rate), to realize separation. There is no selectivity on the raw gas for membrane itself. The membrane only provides isolation for gas and absorption liquid. The gas and the liquid is not direct contact. They flow on both sides of the membrane (Gao et al., 2016). the absorbents play a key role in the selectivity for the composition. The membrane absorption method has overcome some weaknesses of traditional gas (or CO_2) absorption methods. In the case of low CO_2 partial pressure, this method has high selectivity and large separation driving force, which can be modularized for engineering design (Zhang & Shu, 2015).

Compared with a traditional tower absorber, membrane-absorption method can increase the contact area of gas and absorption solution, and overcome the entrainment phenomenon caused by the direct contact between two-phase of gas–liquid, it has the advantages, including stable mass transfer interface, large specific surface area, high mass transfer efficiency, low energy consumption, no bubble, no flooding, small equipment size, higher operation flexibility and easy maintenance (Gao et al., 2016).

Regarding the demonstration of membrane technologies, there was few demonstration projects had been carried out. Lie J. A. et al. tested the performance of three different types of membranes in capturing CO_2 from the blast furnace gas (BFG). Australia CO_2CRC project developed a new membrane separation technology, which is applied to separate CO_2 for industry gases (CO_2CRC).

2.5 Cryogenic Distillation Capture Technology

Cryogenic separation technology refers to compression, liquefaction and distillation of a gas mixture at a low temperature and high pressure operating condition, according to different gas composition with different boiling points. CO_2 could be removed and purified from a mixture gas under a certain temperature and pressure condition. The purity of extracted CO_2 by this method is high, but it takes a lot of extra energy in the liquefaction and compression process (Xiao et al., 2012). In nature, the technology is mainly a gas liquefaction process. Normally, a flue gas is compressed and cooled until a phase change of CO_2, to achieve distillation purpose of separating CO_2 from the flue gas. The advantage of cryogenic separation technology is that the technology can produce liquid CO_2 directly for further transport by truck vehicle, thus avoid further compression and liquefaction of gaseous CO_2 product; the disadvantage of such a technology are in two aspects: one is large energy consumption for cryogenic process, especially for low CO_2 concentration gas stream; the second is blocking issue by water, which will become solid under low temperature. The technology could be more economical for high concentration of CO_2 (above 60%) (Wang, 2007).

The process flow of a cryogenic distillation capture unit is mainly divided into three parts: compression and cooling of feed gas, rectification separation and cryogenic distillation. Practically, a CO_2 enriched feed gas stream is firstly compressed and cooled to about 4.4 Mpa and 40 °C, then sent to a cold tank for rectification purification. In the rectification tower, the light component is gradually gathered at the top of the tower, CO_2 is gradually gathered in the tower bottom at high concentration. Finally, the high concentration CO_2 is super cooled to − 25 °C and put into storage tanks with a liquid phase after throttling. The liquid CO_2 is then delivered to a CO_2 tank by charging pump, and transported to storage sites or utilization facilities by a vehicle. In the project of top gas recycle of blast furnace (TGRBF) under ULCOS program, CO_2 was captured by pressure swing adsorption (PSA) method, followed by a cryogenic distillation technology to generate liquefied CO_2 (Birat, 2010).

2.6 Summary

For steel/iron industry, carbon capture technology is not only a large-scale CO_2 emission reduction technology with great potential, but also the only technology that can facilitate a large-scale direct emission reduction for steel/iron. It can build the capacity of large-scale enterprises in dealing with sustainable emission reduction without changing their own energy balance system.

A large portion of CO_2 associated with steel/iron production can be sourced from post-combustion flue gas. Correspondingly, post-combustion capture technologies, including absorption separation technology, adsorption separation technology, membrane separation technology, and cryogenic distillation technology can be suitable. Overall, among these technologies, the absorption method is the most mature. In particular, chemical absorption method has already been demonstrated by two large-scale commercial demonstration projects with capacity of one million tons per annum CO_2 for each in coal-fired power generation industry.

The physical absorption method is demonstrated by commercial application in chemical industry very early. It is relatively mature in carbon capture applications. The membrane separation method is also widely used in gas separation industry. In a steel/iron production process, different kinds of capture technologies have different technical applicability and economic performance for different gas sources, as different emission sources have different characteristics. It is necessary to provide an insightful analysis on the technical applicability and economic performance of each capture method for each gas source based on different capture requirements.

3 Section 3

3.1 Techno-Economic Analysis of CO_2 Capture from Different Sources

Absorption separation, adsorption separation, membrane separation, cryogenic distillation separation and other capture technologies all have their own features in terms of technology applicability and economic performance. Table 7 summa-

Table 7 Characteristics of CO_2 separation technologies

Technology	Suitable condition		Strength	Weakness
Absorption	Chemical method	Low CO_2 concentration gas source	High CO_2 recovery rate; commercial demonstration available	Chemical process, produce secondary pollution
	Physical method	High CO_2 concentration gas source	High separation efficiency, low energy consumption	high content CO_2 required
Adsorption	Pressure swing adsorption	High CO_2 concentration gas source	Less corrosion and environmental pollution, high CO_2 recovery rate	Large land use, complex system
	Temperature swing adsorption	High CO_2 concentration gas source	Less corrosion and environmental pollution, high CO_2 recovery rate	Slow temperature change, large land use, complex system

(continued)

Table 7 (continued)

Technology	Suitable condition	Strength	Weakness
Membrane separation	High CO_2 concentration gas source	Small land use, simple system, easy operation and maintenance	High energy consumption, easy to be contaminated with particulate matter
Cryogenic distillation	High CO_2 concentration gas source	Liquefaction CO_2 can be obtained directly	High energy consumption, high investment, products can only be liquid

rized the overall strength and weakness of each technology (Li, 2014; Han, 2009; Wang, 2016; Huang, 2008). In general, the CO_2 concentration of each emission source in a steel/iron production process is low, thus physical absorption method is obviously unsuitable for its own use. a temperature swing adsorption method may not be suitable for large scale carbon capture because of a slow temperature change. As indicated in Sect. 3, because the composition, pressure and temperature properties of different emission sources in a steel production process are different, the technical and economic may be different when applied to different emission sources.

In China, CO_2 are classified into four different quality specifications, i.e. food grade, industrial grade, internal use, and storage grade. For a concise analysis in this study, industrial grade and food grade are grouped into an industrial grade above category, with CO_2 concentration being greater than 99.9%. The internal use require that CO_2 concentration be greater than 95% and O_2, H_2, and CO be less than 0.5%; while storage grade only expects the 95% of CO_2. Capture scale can be divided into small scale (10,000 tons/year) and large scale (500,000 tons/year).

The captured CO_2 has three different use purposes: internal use, large-scale external use by pipeline transportation and small-scale external use by tanker transportation. The internal use only needs to be pressurized to 0.6Mpa to be delivered directly by an internal pipeline; the large-scale external use by pipeline transportation requires the CO_2 be transported to designated locations on large scale in a supercritical/dense or a gaseous phase; the small scale external use by tanker requires the captured CO_2 been transported in a liquid phase. Both different specification and different use requirements affect technical and economic assessment of carbon capture technologies.

In view of multiple options of capture methods, the technical assessment of a capture method is mainly measured from the perspective of applicability and maturity in CO_2 capture field. The criterion can be benchmarked by three levels: good, average and bad. The economic assessment is mainly measured based on current capital cost and operating cost and potential capital cost and operating cost by year 2025, when is the end year of phase 1 of the Plan of China's Coal Clean and High Efficiency Utilization Key Project in Facing 2030 Key Project Program (under developing). The operation time assumed to be 8000 h/year and the final CO_2 product is in as gaseous at an atmospheric pressure and temperature of 25 °C.

3.2 Techno-Economic Analysis of CO_2 Capture from Blast Furnace Gas

Taking a blast furnace gas (BFG) from a Chinese steel/iron mill as a case (Table 8), the content of combustible component CO accounts for about 21.8%, the content of CO_2 and N_2 are 23.1% and 51.7% respectively. Pressure is 206 kPa (gauge pressure) and temperature is 150 °C. The particulate matter concentration in the gas stream is high containing saturated water moisture. The flue gas adopts a TRT system to recover energy in blast furnace gas. After TRT, pressure of the blast furnace gas is slightly higher than an atmospheric pressure, and the temperature is cooled to room temperature. The content of water is also reduced to a saturated level corresponding to the temperature and pressure. Therefore, there are two different gas sources for capture CO_2 from the blast furnace gas, i.e., flue gases before TRT and after TRT, respectively.

3.2.1 Technical Performance

After dust removal and purification processes, the blast furnace gas before TRT still has a temperature of 100 °C and a pressure of 0.2–0.3 Mpa. If the purity of the captured CO_2 is required to be industrial grade above for both internal and external uses, chemical absorption method is well technical suitable as it has advantages of high concentration of CO_2 capture through selective absorption and desorption, easy removal of other impurities, and proved by large-scale demonstration project in power-generation industry. It is suitable for the capture of large-scale low concentration CO_2, but the capture process involves chemical process which is complex to iron/steel industry, and the safety and environmental protection requirements for use and storage of absorbent are high. For internal use of small-scale capture, the absorption method is also applicable.

The adsorption method can also be used to produce large-scale high concentration CO_2, however, requiring large land space. The small-scale capture by adsorption is more applicable, due to reduced land use space. In contrast, the cryogenic method is not applicable for CO_2 capture for internal use, because the internal use CO_2 can be delivered through an internal pipeline without liquefaction. With respect to the membrane separation method, it is difficult to guarantee

Table 8 Composition of typical blast furnace gas in China

Composition	CO_2	CO	H_2	N_2	O_2	H_2O
Vol %	23.1	21.8	3.2	51.7	0.2	Saturation

a CO_2 product quality of industrial grade above by membrane technology itself. hence, the applicability of technology only by membrane is poor. However, the combination of membrane separation method and adsorption method (PSA) can overcome the disadvantage of adsorption method (large land use), while ensuring a high quality product by adsorption method. Similarly, the integration of membrane separation with chemical absorption can also ensure a high quality CO_2 product, but it is challenged by complex of membrane separation and chemical reaction process in one system.

For external use of CO_2 with industrial grade above, the demand is normally small. The captured CO_2 needs to be pressurized and liquefied, and temporally stored in a high pressure and low-temperature storage tank. Then, it is transported to the designated site by a truck. Under this circumstance, except chemical absorption and pressure swing adsorption methods, cryogenic method can also be applied. The cryogenic method that combined with membrane separation or pressure swing adsorption can obtain liquefied CO_2 products directly, leading to the improvement of applicability.

Overall, for internal use of food grade or industrial grade above CO_2, absorption method is a mature technology, thus more applicable than other methods without consideration of its environmental impacts; otherwise, the combination of membrane and adsorption technology is more suitable. For external use of food grade or industrial grade CO_2, the scale of project is usually small as market demand is limited. As truck transportation is preferred, membrane separation or pressure swing adsorption combined with cryogenic distillation is also applicable.

In the situation that the captured CO_2 merely needs to meet requirements for general storage or internal use purposes (CO_2 concentration is greater than 95%) (Table 10), chemical absorption method is applicable, but the advantages are less obvious than other technologies. The applicability of pressure swing adsorption method will be improved due to the reduced requirements of CO_2 product, especially for smaller scale, the disadvantage of land use is not so obvious. The use of cryogenic method depends mainly on whether the final CO_2 product needs to be liquefied. Because the concentration requirement of CO_2 product is low, membrane separation method becomes applicable. Through the integration of membrane separation and pressure swing adsorption methods, the weakness of large land use of pressure swing adsorption method will be overcome; integration of membrane and chemical absorption method can further enhance the applicability of absorption method, but the technology is still at early demonstration stage.

After the pressure energy of blast furnace gas is recovered by TRT, gas pressure drops slightly to environmental pressure level and temperature decreases to room temperature. Although the change of pressure will affect absorption, adsorption and membrane separation processes more or less. The effect on adsorption and membrane separation methods are slightly greater than that on absorption method, because the required pressure by adsorption and membrane separation is much higher than that of absorption method. Comparing before and after TRT, for membrane separation and pressure swing adsorption method, the main difference is that the investment of a compressor can be saved in the process as pressure of feed gas before TRT is high. Correspondingly, membrane separation and pressure swing adsorption methods are more suitable before TRT than after TRT.

To summarize, under low concentration requirement of CO_2 products, the applicability of adsorption and membrane separation methods has been improved effectively, especially that of the membrane separation method as its weakness of producing high concentration CO_2 is no longer exist. Further, if membrane separation and pressure swing adsorption are integrated, the applicability of the integrated technology may exceed the absorption method by year 2025, especially in the situation of small-scale CO_2 capture project (Tables 9 and 10).

3.2.2 Economic Analysis

(a) **Cost analysis**

Because gas temperature and pressure are different before TRT and after TRT, the economic performance of each capture technology is also different before and after TRT. Under the case of capture rate of 90%, Tables 11 and 12 respectively illustrate the economics performance of capture technologies for CO_2 products with industrial grade above and storage grade from blast furnace gas before TRT.

For large-scale industrial grade CO_2 before TRT, the economics of chemical absorption method is the best with current technology, especially the investment cost is significantly lower than that of other methods. However, investment cost of membrane separation method will decrease relatively quickly as it is under a quick development in China. It is estimated that by 2025, the investment cost of membrane-pressure swing adsorption integration method might be close to that of absorption method, due to a pressure of 0.2–0.3 Mpa before TRT can be utilized; meanwhile, its operating cost has obvious advantages over chemical absorption method. In the case of internal use only need 0.6 Mpa compression pressure for CO_2 transportation, it is expected that by 2025, the cost of membrane-pressure swing adsorption method for blast furnace gas could be around $25–27. With the development of membrane absorption technology, its economic performance could keep up with that of chemical absorption method.

Table 9 Technical applicability of technologies for industrial grade CO_2 from BFG prior to TRT

Product requirement	Technology options	Applicability	Advantage	Disadvantage
Internal use	Chemical absorption	Good	Easy to produce high concentration of CO_2, mature technology, less energy loss in feed gas	Involves a chemical process, the system is complex, negative effect on environment
	Pressure swing adsorption	Average	Can produce high concentration of CO_2, can use pressure of feed gas before TRT, easy to be modular developed	High pretreatment requirement, large land use and unit size, complex system
	Membrane separation	Bad	N/A	Difficult to obtain high purity CO_2 products alone
	Cryogenic distillation	Bad	N/A	Product of is in liquid state, not suitable for internal use
	Membrane—absorption integration	Average	Can guarantee high concentration of CO_2; can reduce land use	Involves integration of membrane and chemical absorption; complex system, and negative impacts on environment; immature technology
	Membrane—adsorption integration	Good	Can produce high concentration CO_2; reduced land use and facility size; enhanced system stability	Involves two processes of membrane and PSA and their integration; system is complex
	Membrane—cryogenic integration	Bad	N/A	Product of is in liquid state, not suitable for internal use

(continued)

Table 9 (continued)

Product requirement	Technology options	Applicability	Advantage	Disadvantage
External use	Chemical absorption	Good	Easy to produce high concentration of CO_2, mature technology, less energy loss in feed gas	Involves a chemical process, the system is complex, negative effect on environment
	Pressure swing adsorption	Average	Can produce high concentration of CO_2, can use pressure of BFG before TRT, easy to be modular developed; reduced land use and unit size	High pretreatment requirement, complex system
	Membrane separation	Bad	N/A	Difficult to obtain high purity CO_2 products alone
	Cryogenic distillation	Average	Product in liquid state, it is suitable for tanker transportation	Large land use and unit size
	Membrane—absorption integration	Average	Can guarantee high concentration of CO_2; can reduce land use	Involves integration of membrane and chemical absorption; complex system, and negative impacts on environment; immature technology
	Membrane—adsorption integration	Good	Can produce high concentration CO_2; reduced land use and facility size; enhanced system stability	Involves two processes of membrane and PSA and their integration; system is complex
	Membrane—cryogenic integration	Good	Product in liquid state, suitable for truck transportation	Involves two processes of membrane and cryogenic separation and their integration; system is complex

Table 10 Technical applicability of technologies for storage Grade CO_2 from BFG prior to TRT

Product requirement	Technology options	Applicability	Advantage	Disadvantage
Internal use	Chemical absorption	Good	Easy to produce high concentration of CO_2, mature technology, less energy loss in feed gas	Involves a chemical process, the system is complex, negative effect on environment
	Pressure swing adsorption	Average	Can produce high concentration of CO_2, can use pressure of feed gas before TRT, easy to be modular developed	High pretreatment requirement, large land use and unit size, complex system
	Membrane separation	Good	Can produce storage grade CO_2, land use is small; system is simple; easy to modular development	Lack of project demonstration, rely on membrane development
	Cryogenic distillation	Bad	Product of is in liquid state, not suitable for internal use	N/A
	Membrane—absorption integration	Average	Can guarantee high concentration of CO_2; can reduce land use	Involves integration of membrane and chemical absorption; complex system, and negative impacts on environment; immature technology
	Membrane—adsorption integration	Good	Can produce high concentration CO_2; reduced land use and facility size; enhanced system stability	Involves two processes of membrane and PSA and their integration; system is complex

(continued)

Table 10 (continued)

Product requirement	Technology options	Applicability	Advantage	Disadvantage
	Membrane—cryogenic integration	Bad	Product of is in liquid state, not suitable for internal use	N/A
External use	Chemical absorption	Good	Easy to produce high concentration of CO_2, mature technology, less energy loss in feed gas	Involves a chemical process, the system is complex, negative effect on environment
	Pressure swing adsorption	Good	Can produce high concentration of CO_2, can use pressure of BFG before TRT, easy to be modular developed; reduced land use and unit size	High pretreatment requirement, complex system
	Membrane separation	Good	Can produce storage grade CO_2, land use is small; system is simple; easy to modular development	Lack of project demonstration, rely on membrane development
	Cryogenic distillation	Average	Product in liquid state, it is suitable for tanker transportation	Large land use and unit size
	Membrane—absorption integration	Average	Can guarantee high concentration of CO_2; can reduce land use	Involves integration of membrane and chemical absorption; complex system, and negative impacts on environment; immature technology

(continued)

Table 10 (continued)

Product requirement	Technology options	Applicability	Advantage	Disadvantage
	Membrane—adsorption integration	Good	Can produce high concentration CO_2; reduced land use and facility size; enhanced system stability	Involves two processes of membrane and PSA and their integration; system is complex
	Membrane—cryogenic integration	Good	Product in liquid state, suitable for truck transportation	Involves two processes of membrane and cryogenic separation and their integration; system is complex

Table 11 Economic performance of technologies for industrial grade above CO_2 capture from BFG before TRT (at capture rate of 90%)

Scale	Technology options	Investment cost (million $)	Operating cost ($/tonne)	Investment cost by 2025 (million $)	Operating cost by 2025 ($/tonne)
Large scale (1.0 mtpa)	Chemical absorption	90–110	30–33	85–105	28–31
	PSA	110–130	29–31	90–110	27–29
	Membrane—absorption	100–120	30–33	90–110	28–31
	Membrane-PSAintegration	120–140	28–30	90–110	25–27
Small scale (0.1 mtpa)	Chemical absorption	11–13	30–33	10–12	28–31
	PSA	12–14	29–31	11–13	27–29
	Membrane—absorption	11–13	30–33	10–12	28–31
	Membrane-PSAintegration	13–15	28–30	10–12	25–27

For storage grade CO_2, compare to other methods, the advantages of chemical absorption method will be lessened. CO_2 purity requirement has little influence on chemical absorption method in terms of either investment cost or operating cost; while pressure swing adsorption method and membrane separation method are sensitive to product requirements. It is very difficult especially for membrane

Table 12 Economic performance of capture CO_2 of storage grade from BFG before TRT (at 90% capture rate)

Scale	Technology options	Investment cost (million $)	Operating cost ($/tonne)	Investment cost by 2025 (million $)	Operating cost by 2025 ($/tonne)	Main changes compared with industrial grade above before TRT
Large scale (1.0 mtpa)	Chemical absorption	90–110	30–33	85–105	28–31	Investment and operation costs of chemical absorption method are not sensitive to the requirement of CO_2 products
	PSA	105–125	28–30	100–120	26–28	Decrease of CO_2 product concentration requirement, lead to the reduction of investment and operating costs

(continued)

Table 12 (continued)

Scale	Technology options	Investment cost (million $)	Operating cost ($/tonne)	Investment cost by 2025 (million $)	Operating cost by 2025 ($/tonne)	Main changes compared with industrial grade above before TRT
	Membrane separation	115–135	30–32	100–125	26–28	Membrane separation technology become feasible with reduced CO_2 product requirement; investment cost and operating cost of the membrane reduce significantly
	Membrane absorption	100–120	30–33	90–110	28–31	Investment and operation costs of chemical absorption method are not sensitive to the requirement of CO_2 products

(continued)

Table 12 (continued)

Scale	Technology options	Investment cost (million \$)	Operating cost (\$/tonne)	Investment cost by 2025 (million \$)	Operating cost by 2025 (\$/tonne)	Main changes compared with industrial grade above before TRT
	Membrane—PSA integration	110–130	27–29	80–100	24–26	Decrease of CO_2 product concentration requirement, lead to a significant reduction of investment and operating costs of membrane separation—PSA
Small scale (0.1 mtpa)	Chemical absorption	11–13	30–33	10–12	28–31	Investment and operation costs of chemical absorption method are not sensitive to the requirement of CO_2 products

(continued)

Table 12 (continued)

Scale	Technology options	Investment cost (million $)	Operating cost ($/tonne)	Investment cost by 2025 (million $)	Operating cost by 2025 ($/tonne)	Main changes compared with industrial grade above before TRT
	Pressure swing adsorption	11.5–13.5	28–30	10.5–12.5	26–28	Decrease of CO_2 product concentration requirement, lead to the reduction of investment and operating costs
	Membrane separation	12–14	30–32	10.5–12.5	26–28	Membrane separation technology become feasible with reduced CO_2 product requirement; investment cost and operating cost of the membrane reduce significantly

(continued)

Table 12 (continued)

Scale	Technology options	Investment cost (million \$)	Operating cost (\$/tonne)	Investment cost by 2025 (million \$)	Operating cost by 2025 (\$/tonne)	Main changes compared with industrial grade above before TRT
	Membrane absorption	11–13	30–33	10–12	28–31	Investment and operation costs of chemical absorption method are not sensitive to the requirement of CO_2 products
	Membrane separation—pressure swing adsorption	12–14	27–29	9.5–11.5	24–26	Decrease of CO_2 product concentration requirement, lead to a significant reduction of investment and operating costs of membrane separation—PSA

separation method to satisfy technical requirements of high concentration CO_2. When the purity requirement is reduced, it is technically feasible for membrane separation method. With the funding support of "membrane method capture CO_2 technology and industrial demonstration project" under the "coal clean and efficient utilization and new energy conversation technology" by Ministry of Science and Technology of China in 2017, it is believed that R&D of membrane technology would be accelerated in the next 5 years. It is expected that by 2025, the investment cost of large-scale capture by membrane technology may be just a little bit higher than that of chemical absorption method. The investment cost and operating cost of membrane integrated with pressure swing adsorption technology would be highly possible less than that of chemical absorption method.

Before TRT, as the feed gas contains high pressure, the advantage of this high pressure on membrane separation and pressure swing adsorption method is far superior to the chemical absorption method. One direct benefit is that a large-scale compressor for pressurized feed gas can be saved; in the case of after TRT, the investment of the compressor is necessary. The investment cost of a compressor is considerable due to the high volume of feed gas (Table 21).

Accordingly, the investment cost of pressure swing adsorption method, membrane separation method and membrane-pressure swing adsorption method after TRT is much higher than that of chemical absorption method. Even if after 2025, the investment cost of membrane separation and pressure swing adsorption methods after TRT is still significantly greater than that of chemical absorption method. Nevertheless, for the operating cost of membrane separation and pressure swing adsorption method, in view that the pressure energy in non-CO_2 airflow can be recycled through the existing TRT system, their operating cost won't be in disadvantageous position compared to chemical absorption method, especially after 2025.

For small scale capture of 100,000 tons/year, the investment cost of membrane separation and pressure swing adsorption technologies before and after TRT are different. The main reason is a high cost associated with a compressor after TRT. The difference of investment cost for chemical absorption method before and after TRT is very small. Hence, for small scale capture after TRT, the economic performance of membrane separation and pressure swing adsorption is far poorer than that of chemical absorption method; while the economics before TRT is just a little inferior to that of chemical absorption method (Tables 13 and 14).

Table 13 Economic performance of technologies for industrial grade CO_2 capture from BFG after TRT (capture rate of 90%)

Scale	Technology options	Investment cost (million $)	Operating cost ($/tonne)	Investment cost by 2025 (million $)	Operating cost by 2025 ($/tonne)	Main changes between prior to TRT and after TRT
Large scale (1.0 mtpa)	Chemical absorption	90–110	30–33	85–105	28–31	Investment and operating costs are unchanged
	PSA	125–145	30–32	105–115	28–30	Compressor is added, leading to rise of investment and operating costs significantly
	Membrane—absorption	90–110	30–33	85–105	28–31	Investment and operating costs are unchanged
	Membrane-PSA integration	135–155	29–31	115–135	26–28	Compressor is added, leading to rise of investment and operating costs significantly
Small scale (0.1 mtpa)	Chemical absorption	11–13	30–33	10–12	28–31	Investment and operating costs are unchanged

(continued)

Table 13 (continued)

Scale	Technology options	Investment cost (million $)	Operating cost ($/tonne)	Investment cost by 2025 (million $)	Operating cost by 2025 ($/tonne)	Main changes between prior to TRT and after TRT
	PSA	14.5–16.5	30–32	13.5–15.5	28–30	Compressor is added, leading to rise of investment and operating costs significantly
	Membrane—absorption	11–13	30–33	10–12	28–31	Investment and operating costs are unchanged
	Membrane-PSA integration	15.5–17.5	29–31	12.5–14.5	26–28	Compressor is added, leading to rise of investment and operating costs significantly

Table 14 Economic of capture CO_2 of storage grade from BFG before TRT (90% capture rate)

Scale	Technology options	Investment cost (million $)	Operating cost ($/ton)	Investment cost by 2025 (million $)	Operating cost by 2025 ($/ton)	Main changes between prior to and after TRT
Large scale (1.0 mtpa)	Chemical absorption	90–110	30–33	85–105	28–31	Investment and operating costs are unchanged
	Pressure swing adsorption	130–150	29–31	110–130	27–29	Compressor is added, leading to rise of investment and operating costs significantly
	Membrane separation	130–150	31–33	115–135	27–29	Compressor is added, the investment cost and operating cost rise significantly
	Membrane absorption	100–120	30–33	90–110	28–31	Investment cost and operating cost are unchanged
	Membrane separation—pressure swing adsorption	125–145	28–30	95–105	25–27	Compressor is added, leading to rise of investment and operating costs significantly

(continued)

Table 14 (continued)

Scale	Technology options	Investment cost (million $)	Operating cost ($/ton)	Investment cost by 2025 (million $)	Operating cost by 2025 ($/ton)	Main changes between prior to and after TRT
Small scale (0.1 mtpa)	Chemical absorption	11–13	30–33	10–12	28–31	Investment and operating costs are unchanged
	Pressure swing adsorption	14–16	29–31	13–15	27–29	Compressor is added, leading to rise of investment and operating costs significantly
	Membrane separation	14.5–16.5	31–33	12.5–14.5	27–29	Compressor is added, leading to rise of investment and operating costs significantly
	Membrane absorption	11–13	30–33	10–12	28–31	Investment cost and operating cost are unchanged
	Membrane separation—pressure swing adsorption	14.5–16.5	28–30	12–14	25–27	Compressor is added, leading to rise of investment and operating costs significantly

(b) Benefit analysis

On the one hand, capturing CO_2 from blast furnace gas (BFG) can reduce emission CO_2, while the captured CO_2 can be used to place nitrogen in iron/steel making processes; on the other hand, removal of CO_2 from BFG will increase the calorific value of BFG. In case that CO_2 is used to replace nitrogen. Supposing one cubic meter of CO_2 is equivalent to one cubic meter of nitrogen, the benefit of one ton CO_2 is about US$15.0/ton when internal price of nitrogen is US$0.03/m^3. If the captured CO_2 can be sold to oil companies for enhancing oil recovery, the price of CO_2 can reach US$20/ton in China. In fact, when CO_2 concentration meets industrial grade or food grade requirements, market sale price can be as high as US$60.0/ ton.

Because of its low calorific value of raw BFG, before CO_2 removal, BFG needs to be mixed with coke oven gas which has high calorific value. Take a mixing ratio of 9:1 as an example, (i.e., nine blast furnace gases mix with one coke oven gas). Based on a case of Chinese iron/steel plant, the unit price of blast furnace gas (BFG) and coke oven gas (COG) is assumed to be US$0.02/Nm^3, respectively. Consequently, the unit price of original gas mixture will be US$0.046/Nm^3. The benefit of blast furnace gas with CO_2 removal is mainly due to the rise of its calorific value in response to the reduced use of coke oven gas. Under different capture rate, the extent of CO_2 removal is different, thus the extent of calorific value increase is also different. Accordingly, the ratio requirement between blast furnace gas and coke oven gas is also different, leading to the cost of mixed gas is also different. The unit cost and benefit of mixed gas with CO_2 removal are shown in Table 15. It is indicated that the higher the capture rate, the more the calorific value increases, the higher the benefit.

Although more CO_2 can be removed with high capture rate, resulting in the rise of the combustion calorific value of blast furnace gas, both investment cost and operating cost of capture technologies may be more expensive under high capture rate. For chemical absorption method, it is easy to achieve 90% capture rate, while for membrane separation and PSA methods, when capture rate reaches 90%, the investment cost and operating cost could be very high. In case that the capture rate is reduced to 85%, the investment cost and operating cost of membrane separation and PSA methods could be significantly reduced. Therefore, it is necessary to comprehensively justify the costs and the benefits of CO_2 based on a real specific business case,

Table 15 Caloric value benefit of CO_2 capture from BFG under different capture rate

Item	Capture rate (%)	BFG proportion	COG proportion	Original cost of mixed gas at ratio of 9:1 of BFG and COG (US$/Nm^3)	Unit cost of mixed gas after CO_2 removal (US$/Nm^3)	Benefit of CO_2 capture from BFG (US$/Nm^3)
1	100	0.988	0.012	0.046	0.023	0.023
2	95	0.984	0.016	0.046	0.024	0.022
3	90	0.98	0.02	0.046	0.025	0.021
4	85	0.976	0.024	0.046	0.026	0.020
5	80	0.973	0.027	0.046	0.027	0.019
6	70	0.966	0.034	0.046	0.029	0.017
7	60	0.959	0.041	0.046	0.031	0.15

3.3 Techno-Economic Analysis of CO_2 Capture from Lime Kiln Flue Gas

3.3.1 Technical Performance

The CO_2 in lime kiln flue gas is directly sourced from the calcination of lime. The composition of lime kiln flue gas is similar to that of coal-fired power generation flue gas. CO_2 concentration is about 15–20% in lime kiln flue gas, which is only a little higher than 11–15% concentration of flue gas of coal-fired power plant, therefore, most of the methods and processes of post-combustion CO_2 capture used by coal-fired power plant basically can be used as a reference to CO_2 capture from lime kiln flue gas by iron/steel industry.

Tables 16 and 17 show the technical performance of capture methods for lime kiln flue gas. Compared to blast furnace gas, the CO_2 concentration of lime kiln flue gas is slightly lower; thus, applicability of chemical absorption method for capturing industrial grade above CO_2 is further improved compared to PSA and membrane separation methods. For the capture of storage grade CO_2, due to the lack of hydrogen and other impurities in lime kiln flue gas, the applicability of membrane separation CO_2 could also be improved, however, the applicability is still at an average level due to the low CO_2 concentration in flue gas. The cryogenic method could be more applicable in case that CO_2 product is required to be transported in liquid phase.

Table 16 Technical performance of capture methods for industrial grade CO_2 from LKFG

Product requirement	Technology options	Applicability	Advantage	Disadvantages
Internal use	Chemical absorption	Good	Capable of producing high concentration CO_2, mature technology	Involves chemical process, complex system and negative effect on environment
	Pressure swing adsorption	Average	Can produce high concentration of CO_2; can be modularly developed	High pretreatment requirement, large land use and unit size, complex system
	Membrane separation	Bad	N/A	Poor system reliability, lack of technology demonstration; difficulty to produce high concentration CO_2
	Cryogenic distillation	Bad	N/A	Product of is in liquid state, not suitable for internal use
	Membrane—absorption integration	Average	Can guarantee high concentration of CO_2; can reduce land use	Involves integration of membrane and chemical absorption; complex system, and negative impacts on environment; immature technology
	Membrane—adsorption integration	Good	Can produce high concentration CO_2; reduced land use and facility size; enhanced system reliability	Involves two processes of membrane and PSA and their integration; system is complex

(continued)

Table 16 (continued)

Product requirement	Technology options	Applicability	Advantage	Disadvantages
	Membrane—cryogenic integration	Bad	N/A	Product of is in liquid state, not suitable for internal use
External use	Chemical absorption	Good	Capable of producing high concentration CO_2, mature technology	Involves a chemical process, the system is complex, negative effect on environment
	Pressure swing adsorption	Average	Can produce high concentration of CO_2; can be modularly developed	High pretreatment requirement, large land use and unit size, complex system
	Membrane separation	Bad	N/A	Poor system reliability, difficult to obtain high purity CO_2 products alone
	Cryogenic distillation	Bad	N/A	CO_2 concentration is too low
	Membrane—absorption integration	Average	Can guarantee high concentration of CO_2; can reduce land use	Involves integration of membrane and chemical absorption; complex system, and negative impacts on environment; immature technology
	Membrane—adsorption integration	Good	Can produce high concentration CO_2; reduced land use and facility size; enhanced system stability	Involves two processes of membrane and PSA and their integration; system is complex

(continued)

Table 16 (continued)

Product requirement	Technology options	Applicability	Advantage	Disadvantages
	Membrane—cryogenic integration	Good	Product of is in liquid state, suitable for external use by small scale	Not suitable for pipeline transportation in super-critical phase

Compared to blast furnace gas, another difference is that the nitrogen composition in lime kiln feed gas is high. After the pressurized feed gas pass through membrane or adsorption adsorbents, the nitrogen content in the gas stream is increased due to removal of CO_2. Without further pressurized, the gas stream with higher nitrogen content can directly generate high purity nitrogen by other nitrogen separation technologies such as membrane and thus produce nitrogen by-product with low costs. Regarding to chemical absorption method, the separation of nitrogen from gas stream still requires additional pressure and pretreatment. From the perspective of technical maturity, a 100 kilo tonnes per year demonstration project by PSA technology for capturing CO_2 from lime kiln flue gas was carried out by Shougang Jingtang steel plant located in Tangshan, China.

Although there is no demonstration projected reported on chemical absorption method, it can be directly applied to lime kiln flue gas with no doubt as it plenty of experience of post-combustion capture demonstration in coal-fired power plant which can certainly referred. Currently membrane separation method lacks technical demonstration, so the maturity of this technology is poor. However, with the accomplishment of the MOST funded "membrane method capture CO_2 technology and industrial demonstration project" under the "coal clean and efficient utilization and new energy conversation technology" by 2022. It can be reasonably estimated that the maturity of membrane method could have a significant improvement by year 2025.

3.3.2 Economic Analysis

Comparatively, because the CO_2 concentration in lime kiln flue gas is slightly higher than in coal-fired power generation flue gas, investment and operating costs of the corresponding technology are slightly lower than post-combustion CO_2 capture in the coal-fired power plant, while also lower than capture technologies applicable to blast furnace gas capture. When considering nitrogen

Table 17 Technical performance of capture methods for storage grade CO_2 from LKFG

Product requirement	Technology options	Applicability	Advantage	Disadvantages
Internal use	Chemical absorption	Good	Capable of producing high concentration CO_2, mature technology	Involves chemical process, complex system and negative effect on environment
	Pressure swing adsorption	Average	Can produce high concentration of CO_2 and can be modularly developed	High pretreatment requirement, large land use and unit size, complex system
	Membrane separation	Average	Can produce storage grade CO_2 and can be modularly developed	Poor system reliability, lack of technology demonstration
	Cryogenic distillation	Bad	N/A	Product of is in liquid state, not suitable for internal use
	Membrane—absorption integration	Average	Can guarantee high concentration of CO_2; can reduce land use	Involves integration of membrane and chemical absorption; complex system, and negative impacts on environment; immature technology
	Membrane—adsorption integration	Good	Can produce high concentration CO_2; reduced land use and facility size; enhanced system stability	Involves two processes of membrane and PSA and their integration; system is complex
	Membrane—cryogenic integration	Bad	N/A	Product of is in liquid state, not suitable for internal use

(continued)

Table 17 (continued)

Product requirement	Technology options	Applicability	Advantage	Disadvantages
External use	Chemical absorption	Good	Capable of producing high concentration CO_2, mature technology	Involves chemical process, complex system and negative effect on environment
	Pressure swing adsorption	Average	Can produce high concentration of CO_2 and can be modularly developed	High pretreatment requirement, large land use and unit size, complex system
	Membrane separation	Average	Can produce storage grade CO_2 and can be modularly developed	Poor system reliability, lack of technology demonstration
	Cryogenic distillation	Bad	N/A	CO_2 concentration is too low
	Membrane—absorption integration	Average	Can guarantee high concentration of CO_2; can reduce land use	Involves integration of membrane and chemical absorption; complex system, and negative impacts on environment; immature technology
	Membrane—adsorption integration	Good	Can produce high concentration CO_2; reduced land use and facility size; enhanced system stability	Involves two processes of membrane and PSA and their integration; system is complex
	Membrane—cryogenic integration	Good	Product of is in liquid state, suitable for external use by small scale	Not suitable for pipeline transportation in super-critical phase

recovery, the investment cost will increase more. Tables 18 and 19 give the economic performance of CO_2 products by industrial above grade and storage grade, respectively.

For the capture of industrial grade above CO_2, the investment cost of chemical absorption method is far below that of pressure swing adsorption and other methods; even by 2025, it is still the case. For operating cost, chemical absorption method is lower than that of other methods under current technical condition, however, it might be at the same level as other methods by 2025. Because membrane method and pressure swing adsorption method are easy to produce high

Table 18 Economic performance of technologies for industrial grade from LKFG (at capture rate of 90%)

Scale	Technology options	Investment cost (million $)	Operating cost ($/ton)	Investment cost by 2025 (million $)	Operating cost by 2025 ($/ton)
Large scale (1.0 mtpa)	Chemical absorption	100–120	32–35	95–115	30–33
	PSA	140–160	33–35	120–140	30–33
	PSA & N_2 recovery	150–170	36–38	140–160	33–35
	Membrane absorption	100–120	32–35	95–115	30–33
	Membrane-PSA integration	145–165	32–34	120–140	29–31
	Membrane-PSA integration & N_2 recovery	155–175	35–37	130–150	32–34
Small scale (0.1 mtpa)	Chemical absorption	13–15	32–35	13–15	30–33
	PSA	16–18	33–35	14–16	30–33
	PSA & N_2 recovery	17–19	36–38	16–18	33–35
	Membrane absorption	13–15	32–35	13–15	30–33
	Membrane-PSA integration	16–18	32–34	13–15	29–32
	Membrane-PSA integration & N_2 recovery	18–20	35–37	16–18	32–35

Table 19 Economic performance of technologies for storage grade from LKFG (at capture rate of 90%)

Scale	Technology options	Investment cost (million $)	Operating cost ($/ton)	Investment cost by 2025 (million $)	Operating cost by 2025 ($/ton)
Large scale (1.0 mtpa)	Chemical absorption	100–120	32–35	95–115	30–33
	PSA	130–150	32–34	110–130	29–32
	PSA & N_2 recovery	140–160	35–37	130–150	32–34
	Membrane absorption	100–120	32–35	95–115	30–33
	Membrane separation	135–155	33–35	105–125	29–32
	Membrane-& N_2 recovery	145–165	36–38	105–125	32–35
	Membrane-PSA integration	135–155	31–33	110–130	28–30
	Membrane-PSA integration & N_2 recovery	145–165	34–36	120–140	31–33
100,000 tons/year	Chemical absorption	13–15	32–35	13–15	30–33
	pressure swing adsorption	15–17	32–34	13–15	29–32
	Adsorption + N_2 recovery	16–18	35–37	15–17	32–34
	Membrane absorption	13–15	32–35	13–15	30–33
	Membrane separation	15–17	33–35	13–15	29–32
	Membrane separation + N_2 recovery	16–18	36–38	15–17	32–34
	Membrane separation—pressure swing adsorption	15–17	31–33	12–14	28–31
	Membrane separation—pressure swing adsorption + N_2 recovery	17–19	34–36	15–17	31–34

purity nitrogen as a by-product, the economics of membrane and pressure swing adsorption methods can be greatly improved, and can probably be higher than chemical absorption method, when this nitrogen benefit is included, At an internal price of nitrogen being USD0.03/Nm^3, the benefit of nitrogen product could be as high as USD22.0/tonne. Take a lime kiln flue gas with 17% CO_2 and 72% nitrogen as an example, capture one tonne CO_2 can generate about \$53 of nitrogen benefit under scenario of CO_2 capture rate being 90%, and nitrogen capture rate being 80%.

3.4 Techno-Economic Analysis of CO_2 Capture from for Hot Blast Furnace Flue Gas

3.4.1 Technical Analysis

Hot blast furnace is one of the important facilities in the process of iron-making. Although the flue gas produced by a single hot blast furnace is generated intermittently, in the actual production process, multiple hot blast furnaces share one flue gas discharge pipeline; therefore, at the end of the pipeline, the discharge of flue gas is almost continuous. The temperature of hot blast furnace flue gas is usually high, being about 250–300 °C; on the one hand, the high temperature has influence on some carbon capture technologies; on the other hand, the high temperature waste heat can also be recycled to offset the energy consumption of carbon capture.

The composition of a typical hot blast furnace flue gas in Chinese steel/iron plant is shown in Table 20. The CO_2 concentration is 28.52%, which is much higher than that of blast furnace gas, lime kiln flue gas and other major CO_2 emission sources in a steel plant. This high concentration of hot blast flue gas can not only benefit the applicability of chemical absorption method, but also pressure swing adsorption method and membrane separation method (Tables 4, 5, 6, 7, 8, 9, 10, 11, 12, 13, 14 and 15). However, the high temperature of flue gas could affect the chemical absorbent consumption and the use of pressure swing adsorption. The advantages of absorption method and pressure swing adsorption method are able to play well when the temperature is reduced; for the membrane

Table 20 Composition of typical hot blast furnace flue gas in China

Composition	CO_2	N_2	O_2	H_2O	Ash/(mg m^{-3})
Hot blast furnace flue gas	28.52	65.15	2.51	3.82	5

separation method, by using high temperature resistant membrane materials, such as zeolite membrane, it can be operated at high temperature; while for pressure swing adsorption method, the high temperature has a negative effect.

Specifically, capture CO_2 with industrial grade above by large scale for internal use, the chemical absorption method is more applicable, although there is a negative effect of high temperature. Due to the higher CO_2 concentration in feed gas, the applicability of pressure swing adsorption method is improved. Similarly, the applicability of the pressure swing adsorption method integrated with membrane separation will also be more enhanced than that of a simple adsorption method. For the storage grade capture, both absorption and pressure swing adsorption methods can be applied when the temperature is reduced, compared to lime kiln flue gas and blast furnace gas, membrane separation method is more applicable. The applicability of pressure swing adsorption integrated with membrane separation will be further improved.

Similar to lime kiln flue gas, the nitrogen composition of hot blast furnace gas is high. After the pressurized feed gas passes through membrane material or pressure swing adsorption unit, the nitrogen content in the gas stream is increased due to removal of CO_2. Without further pressurized, the gas stream with higher nitrogen content can directly generate high purity nitrogen by other nitrogen separation technologies such as membrane and thus produce nitrogen by-product with low costs. While for the chemical absorption method, the gas stream rich in nitrogen still requires further high energy consumption to separate nitrogen. It is difficult for chemical absorption method itself to obtain high purity nitrogen as a by-product.

From the perspective of technology maturity, the maturity of chemical absorption method is higher than that of membrane separation and pressure swing adsorption. The composition of flue gas from a hot blast furnace contains high concentration CO_2, hence, the experience of industrial gases separation by membrane separation method and pressure swing adsorption method can be referred, although there is still lack of direct technical demonstrations of these two methods (Table 21).

3.4.2 Economic Analysis

The CO_2 concentration in hot blast furnace flue gas is higher than that in blast furnace gas. It is nearly double the CO_2 concentration in lime kiln flue gas. The investment and operating costs of the corresponding capture technologies are far less than that of hot blast furnace gas and blast furnace gas. When considering the recovery of nitrogen, the investment cost will increase. Tables 22 and 23 give the economics of CO_2 product of industrial grade and storage grade, respectively.

Table 21 Technical performance of technologies for hot blast furnace flue gas

Product requirement	Technology	Applicability		Disadvantages
Industrial grade	Chemical absorption	Good	Capable of producing high concentration CO_2, mature technology	Involves chemical process, complex system and negative effect on environment
	pressure swing adsorption	Average	Easy to be modularly developed, can produce by-product nitrogen	High pretreatment requirement, large land use and unit size, complex system
	Membrane separation	Bad	N/A	Poor system reliability, lack of technology demonstration
	Cryogenic distillation	Bad	N/A	Product of is in liquid state, not suitable for internal use
	Membrane—absorption integration	Average	Can guarantee high concentration of CO_2; can reduce land use	Involves integration of membrane and chemical absorption; complex system, and negative impacts on environment; immature technology
	Membrane—adsorption integration	Good	Can produce high concentration CO_2; reduced land use and facility size; enhanced system stability	Involves two processes of membrane and PSA and their integration; system is complex
	Membrane—cryogenic integration	Average	Product is in liquid state, suitable for external use by small scale	Product of is in liquid state, not suitable for internal use

(continued)

Table 21 (continued)

Product requirement	Technology	Applicability		Disadvantages
Storage grade	Chemical absorption	Good	Capable of producing high concentration CO_2, mature technology	Involves chemical process, complex system and negative effect on environment
	Pressure swing adsorption	Good	Easy to be modularly developed, can produce by-product nitrogen	High pretreatment requirement, large land use and unit size, complex system
	Membrane separation	Good	Can be modularly developed, can produce by-product nitrogen	Poor system reliability, lack of technology demonstration
	Cryogenic distillation	Bad	N/A	CO_2 concentration is too low
	Membrane—absorption integration	Average	Can guarantee high concentration of CO_2; can reduce land use	Involves integration of membrane and chemical absorption; complex system, and negative impacts on environment; immature technology
	Membrane—adsorption integration	Good	Can produce high concentration CO_2; reduced land use and facility size; enhanced system stability	Involves two processes of membrane and PSA and their integration; system is complex
	Membrane—cryogenic integration	Average	Product is in liquid state, suitable for external use by small scale	Not suitable for pipeline transportation in super-critical phase

Table 22 Economic performance of technologies for industrial grade from hot blast furnace flue gas (at capture rate of 90%)

Scale	Technology options	Investment cost (million $)	Operating cost ($/tonne)	Investment cost by 2025 (million $)	Operating cost by 2025 ($/tonne)
Large scale (1.0 mtpa)	Chemical absorption	80–100	28–30	75–95	27–29
	PSA	110–130	28–30	100–120	26–28
	PSA & N_2 recovery	125–145	31–33	115–135	29–31
	Membrane absorption	80–100	28–30	75–95	27–29
	Membrane-PSA integration	120–140	27–29	110–130	25–27
	Membrane-PSA integration & N_2 recovery	135–155	33–35	120–140	30–32
Small scale (0.1 mtpa)	Chemical absorption	11–13	28–30	10–11	27–29
	PSA	13–15	28–30	12–14	26–28
	PSA & N_2 recovery	15–17	31–33	14–16	29–32
	Membrane absorption	11–13	28–30	10–11	27–29
	Membrane-PSA integration	14–16	27–29	13–15	25–27
	Membrane-PSA integration & N_2 recovery	16–18	33–35	16–18	30–32

For capture of industrial grade above CO_2, the investment cost of chemical absorption method is still lower than that of PSA and other methods, even by 2025. However, compared with blast furnace gas, this gap is much smaller. The operating cost of PSA method is low. By 2025, the operating cost of pressure swing adsorption method may be lower than that of absorption method. In particular, membrane-PSA integration method is very likely to be lower than absorption method. If considering the benefit of by-product nitrogen, the cost of membrane-PSA integration method can be greatly improved, and will be lower than that of chemical absorption method.

For CO_2 capture of storage grade, the economics of chemical absorption method is slight lower than that of industrial grade capture. The investment

Table 23 Economic performance of technologies for storage grade from hot blast furnace flue gas (at capture rate of 90%)

Scale	Technology options	Investment cost (million $)	Operating cost ($/ton)	Investment cost by 2025 (million $)	Operating cost by 2025 ($/ton)
Large scale (1.0 mtpa)	Chemical absorption	80–100	28–30	75–95	27–29
	PSA	100–120	27–29	90–110	25–27
	PSA & N_2 recovery	110–130	30–32	100–120	28–30
	Membrane absorption	80–100	28–30	75–95	27–29
	Membrane seperation	105–125	28–30	90–110	25–27
	Membrane-& N_2 recovery	115–135	31–33	100–120	28–30
	Membrane-PSA integration	110–130	28–30	95–115	25–27
	Membrane-PSA integration & N_2 recovery	120–140	31–33	105–125	28–30
Small scale (0.1 mtpa)	Chemical absorption	11–13	28–30	10–11	27–29
	PSA	12–14	27–29	11–13	25–27
	PSA & N_2 recovery	14–16	30–32	13–15	28–30
	Membrane absorption	11–13	28–30	10–11	27–29
	Membrane seperation	13–15	28–30	11–13	25–27
	Membrane-& N_2 recovery	15–17	31–33	13–15	28–30
	Membrane-PSA integration	13–15	26–28	12–14	24–26
	Membrane-PSA integration & N_2 recovery	15–17	29–31	14–16	27–29

cost and operating cost of PSA method will be further declined. The cost of membrane separation is also close to the cost of absorption method and PSA method. It may likely be lower than the cost of PSA method in 2025. The cost of membrane-PSA integration method could be further reduced than that of single membrane separation method and single PSA method. At an internal price of nitrogen being USD0.03/Nm^3, the benefit of nitrogen product could be as high as USD22.0/tonne. Take a lime kiln flue gas with 28% CO_2 and 65% nitrogen as an example, capture one tonne CO_2 can generate about $30 of nitrogen benefit under scenario of CO_2 capture rate being 90%, and nitrogen capture rate being 80%. Compared with the capture of lime kiln flue gas, although the cost of CO_2 capture decreases, the benefit of nitrogen also decreases.

3.5 Techno-Economic Analysis of CO_2 Capture for Linz-Donawitz Gas

In a Linz-Donawitz steel-making process, under high temperature, carbon in molten iron reacts with blown oxygen to generate a mixture of carbon monoxide and a small amount of carbon dioxide. The recovered Linz-Donawitz gas (LDG) contains about 60% of CO, 15–20% of CO_2, as well as nitrogen, hydrogen and trace oxygen. Capturing CO_2 from Linz-Donawitz gas is not only the need for carbon reduction, but also anti-corrosion of gas pipe caused by the presence of CO_2. Because the amount of Linz-Donawitz gas is not generated at a stable level in a smelting process, the gas usually needs to be stored in a gas cabinet temporarily after full dusting and cooling. The capacity of the gas cabinet determines the gas buffer time in the gas cabinet.

With the gas cabinet storage, "intermittent producing, continuous use" of the gas can be realized. CO_2 in the gas stream can be captured either before the gas stream enters the cabinet or after the gas stream enters the cabinet. Capturing CO_2 before the cabinet can reduce the volume of the gas stream, and increase the buffer efficiency of a cabinet; on the other hand, the dissolution of CO_2 is reduced at a water sealing process through removing CO_2 in the gas stream, thereby, the corrosivity of the gas stream is further undermined. When capturing CO_2 after the gas cabinet, the corresponding gas volume can be continuous and stable which is particularly helpful to some capture technologies.

3.5.1 Technical Analysis

For capture CO_2 from a LDG stream before a gas cabinet, whatever the captured CO_2 product is at industrial above grade or storage grade, whatever it is by large scale or small scale, overall, the technical applicability of chemical absorption method and membrane-absorption method is obviously better than that of other methods. The main reason is that the intermittence of Linz-Donawitz gas before a cabinet storage has less influence on a chemical absorption process, while greater influence on membrane separation and pressure swing adsorption processes that require compression operation of a compressor. Especially for a centrifugal compressor, a gas stream without stable volume can result in many technical problems (Table 32).

However, when capture CO_2 from LDG stream after a gas cabinet, if capturing CO_2 for internal use, pressure swing adsorption method and membrane separation method can also be applicable since feed gas stream can be adjusted at stable level with the cabinet. Another technical advantage of membrane and PSA separation methods is that the CO_2 removed gas stream can be directly transported through pipeline without further pressure because the feed gas has been pressurized for removal the CO_2. The integration of membrane separation and PSA can further enhance the advantage. If capturing CO_2 for industrial grade above, absorption method is more applicable (Tables 24, 25 and 26).

3.5.2 Economic Analysis

Before a Linz-Donawitz gas enters a cabinet, chemical absorption is cost-effective to capture CO_2 of industrial grade above and storage grade. For the project of million tonnes/year scale (large scale), because of the intermittency of feed gas, the investment cost of such a large-scale project will be greater than that of post-combustion capture project with the same concentration in a coal-fired power plant. The operating cost of a project is only slightly higher than that of the post-combustion capture project in coal-fired power plant, because the costs mostly depend on energy consumption of absorption and desorption processes, which are less affected by the intermittency of feed LDG stream. Removing CO_2 from a Linz-Donawitz gas can not only increase the calorific value of the gas, but also increase the efficiency of gas storage cabinets and reduce the corrosive effect of the gas on the subsequent device and equipment. In addition, the capture process can also contribute partially to cool the gas, thus save the capital and operating costs of cooling equipment.

Regarding the CO_2 from LDG stream after a gas cabinet, a chemical absorption method is also cost effective. The investment cost is further reduced significantly while the operating cost also reduces as the feed gas has a stable volume. However, the feed gas before cabinet has a high corrosivity to the cabinet, and the utilization efficiency of the cabinet is also poor. Currently, the investment cost of PSA and membrane separation methods that are applicable to storage grades

Table 24 Technical performance of technologies for Linz-Donawitz gas before cabinet

Technological options	Technical applicability	Main reasons	Disadvantages
Chemical absorption	Good	Capable of producing high concentration CO_2, mature technology	Involves chemical process, complex system and negative effect on environment
Pressured swing adsorption	Bad	Compressor is difficult to operate under an intermittent condition	High pretreatment requirement, large land use and unit size, complex system
Membrane separation	Bad	Compressor is difficult to operate under an intermittent condition	Poor system reliability, lack of technology demonstration
Cryogenic distillation	Bad	Compressor is difficult to pressurize the intermittent gas	N/A
Membrane—absorption integration	Average	Can guarantee high concentration of CO_2; can reduce land use	Involves integration of membrane and chemical absorption; complex system, and negative impacts on environment; immature technology

are far higher than that of chemical absorption method, however, after 2025, the investment cost could decrease significantly. When considering the energy consumption saved in the subsequent compressing process that ensure the gas transported through a pipeline under high pressure, the operating cost may further decrease. Considering this cost saving, by 2025, the overall cost of the technology that integrates membrane separation with PSA is probably close to the cost of chemical absorption method (Tables 27 and 28).

Table 25 Technical performance of technologies for industry grade CO_2 from Linz-Donawitz gas after cabinet

Product requirement	Technology options	Applicability	Advantage	Disadvantages
Internal use	Chemical absorption	Good	Capable of producing high concentration CO_2, mature technology	Involves chemical process, complex system and negative effect on environment
	Pressured swing adsorption	Average	Can produce high concentration of CO_2 and can be modularly developed	High pretreatment requirement, large land use and unit size, complex system
	Membrane separation	Bad	N/A	Poor system reliability, lack of technology demonstration; difficulty to produce high concentration CO_2
	Cryogenic distillation	Bad	N/A	Product of is in liquid state, not suitable for internal use
	Membrane—absorption integration	Average	Can guarantee high concentration of CO_2; can reduce land use	Involves integration of membrane and chemical absorption; complex system, and negative impacts on environment; immature technology
	Membrane—adsorption integration	Good	Can produce high concentration CO_2; reduced land use and facility size	Involves two processes of membrane and PSA and their integration; system is complex
	Membrane—cryogenic integration	Bad	N/A	Product of is in liquid state, not suitable for internal use

(continued)

Table 25 (continued)

Product requirement	Technology options	Applicability	Advantage	Disadvantages
External use	Chemical absorption	Good	Capable of producing high concentration CO_2, mature technology	Involves a chemical process, the system is complex, negative effect on environment
	Pressure swing adsorption	Average	Can produce high concentration of CO_2 and can be modularly developed	High pretreatment requirement, large land use and unit size, complex system
	Membrane separation	Bad	N/A	Poor system reliability, difficult to obtain high purity CO_2 products alone
	Cryogenic distillation	Bad	N/A	CO_2 concentration is too low
	Membrane—absorption integration	Average	Can guarantee high concentration of CO_2; can reduce land use	Involves integration of membrane and chemical absorption; complex system, and negative impacts on environment; immature technology
	Membrane—adsorption integration	Good	Can produce high concentration CO_2; reduced land use and facility size	Involves two processes of membrane and PSA and their integration; system is complex
	Membrane—cryogenic integration	Good	Product of is in liquid state, suitable for external use by small scale	Not suitable for pipeline transportation in super-critical phase

Table 26 Technical performance of technologies for storage grade CO_2 from Linz-Donawitz Gas after cabinet

Product requirement	Technology options	Applicability	Advantage	Disadvantages
Internal use	Chemical absorption	Good	Capable of producing high concentration CO_2, mature technology	Involves chemical process, complex system and negative effect on environment
	Pressure swing adsorption	Average	Can produce high concentration of CO_2 and can be modularly developed, LDG after CO_2 removal can enter pipeline without addition pressure	High pretreatment requirement, large land use and unit size, complex system
	Membrane separation	Average	Can be modularly developed. LDG after CO_2 removal can enter pipeline without addition pressure	Poor system reliability, lack of technology demonstration
	Cryogenic distillation	Bad	N/A	Product of is in liquid state, not suitable for internal use
	Membrane—absorption integration	Average	Can guarantee high concentration of CO_2; can reduce land use	Involves integration of membrane and chemical absorption; complex system, and negative impacts on environment; immature technology

(continued)

Table 26 (continued)

Product requirement	Technology options	Applicability	Advantage	Disadvantages
	Membrane—adsorption integration	Good	Can produce high concentration CO_2; reduced land use and facility size; enhanced system stability. LDG after CO_2 removal can enter pipeline without addition pressure	Involves two processes of membrane and PSA and their integration; system is complex
	Membrane—cryogenic integration	Bad	N/A	Product of is in liquid state, not suitable for internal use
External use of storage grade	Chemical absorption	Good	Capable of producing high concentration CO_2, mature technology	Involves chemical process, complex system and negative effect on environment
	Pressure swing adsorption	Average	Can produce high concentration of CO_2 and can be modularly developed. LDG after CO_2 removal can enter pipeline without addition pressure	High pretreatment requirement, large land use and unit size, complex system
	Membrane separation	Average	Can be modularly developed; LDG after CO_2 removal can enter pipeline without addition pressure	Poor system reliability, lack of technology demonstration
	Cryogenic distillation	Bad	N/A	CO_2 concentration is too low

(continued)

Table 26 (continued)

Product requirement	Technology options	Applicability	Advantage	Disadvantages
	Membrane—absorption integration	Average	Can guarantee high concentration of CO_2; can reduce land use	Involves integration of membrane and chemical absorption; complex system, and negative impacts on environment; immature technology
	Membrane—adsorption integration	Good	Can produce high concentration CO_2; reduced land use and facility size; enhanced system stability; LDG after CO_2 removal can enter pipeline without addition pressure	Involves two processes of membrane and PSA and their integration; system is complex
	Membrane—cryogenic integration	Good	Product of is in liquid state, suitable for external use by small scale; LDG after CO_2 removal can enter pipeline without addition pressure	Not suitable for pipeline transportation in super-critical phase

Table 27 Economic performance of technologies for industrial grade CO_2 from Linz-Donawitz gas after cabinet (at capture rate of 90%)

Scale	Technology options	Investment cost (million $)	Operating cost ($/tonne)	Investment cost by 2025 (million $)	Operating cost by 2025 ($/tonne)
Large scale (0.1 mtpa)	Chemical absorption	110–130	33–36	100–120	31–34
	adsorption	160–180	35–37	140–160	32–34
	Membrane absorption	130–150	33–36	120–140	31–34
	Membrane separation—pressure swing adsorption	160–180	36–38	140–160	32–34
Small scale (0.1 mtpa)	Chemical absorption	15–17	33–36	14–16	31–34
	Pressure swing adsorption	18–20	35–37	16–18	i
	Membrane absorption	16–18	33–36	15–17	31–34
	Membrane separation—pressure swing adsorption	18–20	36–38	16–18	33–34

3.5.3 Techno-Economic Analysis of CO_2 Capture for Flue Gas from Coke Oven and Power Generation Unit

The coke oven flue gas in a steel/iron plant is sourced from combustion for heating a coke oven. It is the combustion product of fossil fuel in a coke oven and discharged through a coke oven chimney. The flue gas sourced from power generation unit mainly comes from combustion of fuel, such as BFG, COG, coal, natural gas. The sources and the compositions of flue gases from coke oven and power generation unit are similar, with the CO_2 concentration being between 10 and 15%. Therefore, a technology that is applicable to one is roughly applicable to the other.

For capturing storage grade CO_2 from a post-combustion flue gas stream with low concentration CO_2, the chemical absorption method is the most suitable among current available capture technologies. So far, there have already been large-scale post-combustion capture demonstration projects in operation in Canada and US, respectively. The technologies and experiences leant from those two projects can be referred to capture of flue gas from a coke oven and a power generation unit in a steel/iron plant. Although the PSA method and membrane separation method are also suitable for carbon capture of storage grade, however,

Table 28 Economic performance of capture CO_2 of storage grade after Linz-Donawitz Gas after cabinet (90% capture rate)

Scale	Technology options	Investment cost (million $)	Operating cost ($/tonne)	Investment cost by 2025 (million $)	Operating cost by 2025 ($/tonne)
Large scale (0.1 mtpa)	Chemical absorption	110–130	33–36	100–120	31–34
	Adsorption	150–170	34–36	130–150	31–33
	Membrane absorption	110–130	33–36	100–120	31–34
	Membrane separation	150–170	35–37	130–150	32–34
	Membrane separation—pressure swing adsorption	150–170	35–37	130–150	31–33
Small scale (0.1 mtpa)	Chemical absorption	15–17	33–36	14–16	31–34
	pressure swing adsorption	17–19	34–36	15–17	31–33
	Membrane absorption	15–17	33–36	14–16	31–34
	Membrane separation	17–19	35–37	15–17	32–34
	Membrane separation—pressure swing adsorption	17–19	35–37	15–17	32–34

it is still in a technological demonstration stage for flue gas from power-generation processes. Even by 2025, it is estimated that it is difficult for large-scale commercial application mainly because the CO_2 concentration in the flue gas is very low. By 2025, membrane-absorption integration method may become gradually technical mature.

As for economic performance, whatever investment cost or operating cost, chemical absorption method is mostly dominant over other methods, such as PSA method and membrane separation method. By year 2025, membrane-absorption method that integrates chemical absorption with membrane separation would further reduce its investment cost and operating cost. It is likely that its economic is getting better than an absorption method. Membrane separation method and PSA method as well as their integration are unlikely to be economically superior to the chemical absorption method by 2025.

3.6 Discussions

In a typical production process of steel/iron in China, main CO_2 emission sources include: lime kiln flue gas, hot blast furnace gas (HBFG), blast furnace gas (BFG), Linz-Donawitz gas (LDG), and flue gas from coke oven and power generation unit. CO_2 concentration as well as temperature, pressure and impurity content of these emission sources are different; therefore, technical applicability and economic performance of capture technologies are also different when applied to different emission sources. Generally, chemical absorption method is suitable for low concentration gas and PSA, membrane separation and cryogenic distillation methods are suitable for high concentration gas, while membrane separation, PSA and cryogenic methods and their integration are also suitable for gas stream with pressure over a mbient environment level, such as blast furnace gas before TRT.

The different demands on captured CO_2 product also affect technical applicability and economic performance of capture technologies. In general, chemical absorption method is more suitable for obtaining CO_2 of industrial level above than other methods. Membrane separation and PSA methods and their integration are also suitable for storage grade CO_2. capture; while cryogenic distillation method is not suitable for large-scale capture of low CO_2 concentration gas in steel/iron plant, but it is applicable when integrated with membrane separation or PSA method for capture small-scale CO_2 product with liquefaction requirement.

The technical applicability and economic performance of capture technologies are also affected by by-product opportunity associated with the removal of CO_2. For instance, the value of blast furnace gas can be raised by CO_2removal; flue gas of hot blast furnace and lime kiln can produce a by-product of nitrogen; corrosivity of Linz-Donawitz gas can be lessened after CO_2 removal.

The technical applicability and economic performance of capture technologies are also depended on capture rate. Chemical absorption method is less influenced by capture rate than that membrane separation and PSA methods. Under a low CO_2 concentration condition, membrane separation and pressure swing adsorption technologies are difficult to achieve a capture with rate of 90%. Even if it can be achieved through their integration, investment and operation costs would be increased significantly. It is good for membrane separation and PSA methods and their integration to improve their technical applicability and economic performance by acceptance of a capture rate as low as 85%. The optimal capture rate should be identified through a comprehensive analysis of capture cost and benefit in a system framework.

For China's steel/iron industry, take technical applicability and economic performance of various technologies into consideration comprehensively, it is recommended that capture on blast furnace gas, hot blast furnace gas and flue gas from lime kiln be prioritized due to the following main reasons: (1) high concentration of CO_2 in those gas stream; (2) calorific value of BFG can be significantly improved by removal of CO_2; (3) nitrogen gas, which is highly demanded by iron/steel industry, can be obtained as a co-benefit to the capture. Though the CO_2 concentration of lime kiln flue gas is lower than hot blast furnace gas, by-product nitrogen benefit is higher than that gas.

In fact, if taking all major carbon emission sources into account in a whole, carbon capture for iron/steel industry could be another different case, in view of interactions among carbon capture from various sources, For example, high temperature feature of hot blast furnace gas and Linz-Donawitz gas can be used for absorbent regeneration by chemical absorption method; as the result of CO_2 capture, calorific value increasing of blast furnace gas can reduce the amount of feed gas for hot blast furnace heating, coke oven heating and power generation, thus could increase the CO_2 concentration in their flue gases. It is then beneficial to improve the economic performance of further capture from those gases. In short, the technical applicability and economic performance of carbon capture technologies for a steel/iron industry could be affected by multiple factors and their interactions under different conditions, it is necessary to carry out future research with a real-world business case to assess the technical applicability and economic performance of CO_2 capture for China iron/steel industry based on individual gas source and overall gas system, respectively.

4 Section 4

4.1 Concluding Remarks

Carbon capture technology is not only a large-scale CO_2 emission reduction technology with great potential for China's iron/steel industry, but also the only technology that can deliver a large-scale direct emission reduction for steel/iron industry, while it can enable large-scale iron/steel companies to achieve a sustainable emission reduction with less effect on their own established energy balance system. This study provided a comprehensive techno-economic analysis of major CO_2 capture technologies for China's steel/iron industry. Based on a typical iron/steel making process of China's iron/steel industry, main CO_2 emission sources including blast furnace gas, Linz-Donawitz gas, hot blast furnace flue gas, lime kiln flue gas and flue gases of coke oven and power generation within an iron/steel plant were characterized. It was indicated that CO_2 concentration of lime kiln flue gas, hot blast furnace gas, blast furnace gas and Linz-Donawitz gas is higher than that of post-combustion flue gas of normal coal-fired power generation industry; CO_2 concentration of flue gases from coke oven and power generation unit within an iron/steel plant is lower than that a post-combustion flue gas from normal coal-fired power generation industry. In general, hot blast furnace flue gas has the highest CO_2 concentration, blast furnace gas ranks the second, followed by Linz-Donawitz gas and lime kiln flue gas.

Various post-combustion capture methods, including chemical absorption, pressure swine adsorption (PSA), membrane separation, cryogenic distillation methods and their integrations, are identified and analyzed from the perspective of their technical applicability and economic performance based on features of individual gas sources. Overall, chemical absorption is the most mature technology and have been demonstrated by two large-scale to post-combustion capture projects. It is particularly applicable and cost-effective to emission source with low CO_2 concentration. Regarding PSA, membrane separation and cryogenic distillation methods, they are more applicable and cost-effective to high concentration CO_2 source and have excellent technical and economic performance on gas stream with high pressure.

It is recognized by this study, that many factors could affect technical and economic performances of carbon capture technologies. Different temperature, pressure, impurity content and CO_2 concentration of each emission source in steel/iron production process would lead to different technical applicability and economic performance of capture technologies; different demands for CO_2 product also affect technical applicability and economic performance of capture technologies; in addition, co-benefit or by-product associated with CO_2 removal affect technical applicability and economic performance of capture technologies. Specifically, calorific value of blast furnace gas can be significantly improved by CO_2 removal; hot blast furnace flue gas has the highest CO_2 concentration and high concentration nitrogen, lead to the lowest capture cost and an opportunity to produce by-product nitrogen with membrane technology, PSA or membrane-PSA integration technologies; Though the CO_2 concentration of lime kiln flue gas is lower than hot blast furnace gas, by-product nitrogen benefit is higher than that gas.

In view of interactions among carbon capture from various sources, the technical applicability and economic performance of carbon capture technologies for a steel/iron industry could be affected by multiple factors and their interactions under different conditions, it is necessary to carry out future research with a real-world business case to assess the technical applicability and economic performance of CO_2 capture for China iron/steel industry based on individual gas source and overall gas system, respectively.

4.2 A Caveat

Understandably, steel plants differ. This is especially true in China. As this and the previous chapter have discussed, many factors affect the technical and economic performance of CCUS, including but not limited to, temperature, pressure, content impurity, and CO_2 concentration of each emission source. In turn, these varying factors would affect costs. Thus, for actual deployment of CCUS in connection with cost considerations, it goes without saying, it is necessary to conduct real-world cases to assess the technical applicability and economic performance of CO_2 capture in light of local- and firm-specific engineering conditions.

Acronym	**Full spelling**
BF/BOF	Blast furnace/basic oxygen furnace
BFG	Blast furnace gas
CCUS	Carbon capture, utilization and storage
CFG	Controlled flow graph
EAF	Electric arc furnace
HBFG	Hot blast furnace gas
LDG	Linz-Donawitz gas
MEA	Mono ethanol amine
PSA	Pressure swing adsorption
TGRBF	Top gas recycle of blast furnace
TRT	Top gas pressure recovery turbine
VPSA	Vacuum pressure swing adsorption

References

Birat, J. P. (2010). Carbon dioxide (CO_2) capture and storage technology in the iron and steel industry. Carbon dioxide (CO_2) capture, transport and industrial applications. *Developments and innovation in carbon dioxide (CO_2) capture and storage technology* (vol. 1). UK: Woodhead Publishing Ltd.

Cheng, H.-H., Shen, J.-F., & Tan, C.-S. (2010). CO_2 capture from hot stove gas in steel making process. *International Journal of Greenhouse Gas Control, 4*, 525–531.

Cheng, H.-H., Shen, J.-F., & Tan, C.-S. (2010b). CO_2 CRC. Fact sheets-Gas separation membranes. Australia Cooperative Research Centre for Greenhouse Gas Technologies (CO_2CRC)

CO_2CRC. Fact sheets-Gas separation membranes. Australia Cooperative Research Centre for Greenhouse Gas Technologies (CO_2CRC).

Feron, P. H., & Jansen, A. E. (2002). CO_2 separation with polyolefin membrane contactors and dedicated absorption liquids-performances and prospects. *Separation and Purification Technology, 34*(27), 231–242.

Fang, L. (2016). Automatic control of top combustion hot blast stove. *Industrial Furnace, 38*(4).

GCCST. (2016). The global status of CGS, 2016 (Summary Report). Global CCS Institute (GCCST).

Goto, K., Chowdhury, F. A., Kodama, S., et al. (2013). Development of a novel CO_2 capture system with chemical absorption process for the integrated steel works. *World Iron & Steel, 13*(6).

Gielen, D. (2003). CO_2 removal in the iron and steel industry. *Energy Conversion and Managament, 44*, 1027–1037.

IEA. (2010). *Energy technology perspectives 2010—scenarios and strategies to 2050* (p. 650). Paris, France.

Jinlian, W. (2007). *Study on new chemical absorbent and process that absorbs CO_2[D]*. Zhejiang University.

Juan, Z., Qichao, G., Yonghui, L., et al. (2017). A study on increasing the gas recovery rate of converter gas dry purification recovery system. *Metallurgical Power, 3*, 49–51.

Kuramochi, T., Ramírez, A., Turkenburg, W., & Faaij, A. (2011). Techno-economic assessment and comparison of CO_2 capture technologies for industrial processes: Preliminary results for the iron and steel sector. *Energy Procedia, 4*, 1981198–1981208.

Kim, J. Y., Han, K., & Chun, H. D. (2009). CO_2 absorption with low concentration ammonia liquor. *Energy Procedia, 1*, 757–762.

Lie, J. A., Vassbotn, T., Hägg, M.-B., Grainger, D., Kim, T.-J., & Mejdell, T. (2007). Optimization of a membrane process for CO_2 capture in the steelmaking industry. *International Journal of Greenhouse Gas Control, 1*, 309–317.

Mingtan, W., Shenlin, X., & Zitong, X. (2016). Present state and latest development of CO_2 capture technology. *Contemporary Chemical Industry, 45*(5), 1002–1005.

Ming, G., Limin, Z., Xiaoqiang, L., et al. (2016). Research status of CO_2 capture and storage technology (CCS) in the coal-fired power plant. *Gas Purification, 6*, 23–26.

Peng, H. (2013). *Study on adsorption of modified activated to $CO_2$from off-gas in power plant [D]*. Hebei University of Science & Technology.

POSCO. (2013). Carbon report-2013. South Korea.

Rui, R., & Duan, W. (2006). Current situation of CO_2 emission in iron and steel producing and its controlling methods. *Science & Technology Review, 24*(10), 53–56.

Rhee, C. H., Kim, J. Y., Han, K., Ahn, C. K., & Chun, H. D. (2011). Process analysis for ammonia based CO_2 capture in ironmaking industry. *Energy Procedia, 4*, 1486–1493.

Shaodong, C. (2016). *Research on production scheduling in iron and steel scrap remanufacturing under the background of carbon reduction [D]*. Southeast University.

Tseitlin, M. A., Lazutkin, S. E., Styopin, G. M. (1994). A flow-chart for iron making on the basis of 100% usage of process oxygen and hot reducing gases injection. *ISIJ International, 34*, 570–573.

Tobiesen, F. A., Svendsen, H. F., & Mejdell, T. (2007). Modeling of blast furnace CO_2 capture using amine absorbents. *Industrial and Engineering Chemistry Research, 46*, 7811–7819.

Wenmin, J. (2015). *Process simulation and experimental study of the capture carbon dioxide by chemical absorption method [D]*. Zhejiang University.

Weifeng, Z., & Jianhui, S. (2015). On the latest advancements in the membrane-based technologies for CO_2 capture. *Journal of Safety and Environment., 15*(3), 205–211.

Xiaosen, L., & Tao, L. (2009). Present development situation of techniques separating carbon dioxide separation techniques inform flue gas separation. *Modern Chemical Industry, 29*(4), 25.

Xinfeng, S. (2008). *Research and design on CO_2recovery technology of limekiln tail gas [D]*. Xi'an University of Architecture and Technology (in Chinese).

Xiaobing, L. (2014). Research progress on Carbon Dioxide capture and separation technology. *Guangdong Chemical Industry, 41*(5), 115–116.

Yoon, Y. N. S., Jung, S., & Kim, Y. (2011). K2CO3/hindered cyclic amine blend (SEFY-1) as a solvent for CO_2 capture from various industries. *Energy Procedia, 4*, 267–272.

Yonggang, W., Tianhong, Y., Yujie, Z., et al. (2008). Comparison of dry cleaning and recovery technique for converter flue gas with wet one. *Industrial Safety and Environmental Protection, 34*(5), 10–12.

Yongjia, H., Shuli, W., Pengyu, Z., et al. (2009). Current status and advances in CO_2 separation and capture technology. *Natural Gas Industry, 29*(12), 79–82.

Yuansheng, X., Guangquan, L., & Xianqing, Y. (2012). Research progress of carbon dioxide capture technology and mechanism. *Chemical Intermediate, 10*, 1–5.

Zhengping, S. (2009). Recovery and utilization of carbon dioxide in industrial waste gas. *China High-Tech Enterprises, 13*, 88–89.

Zhi, G., Xinmiao, Z., Chenlin, Z., et al. (2016). Research development of membrane materials for separation of CO_2 from flue gas. *Modern Chemical Industry, 6*, 42–45.

斉間 等,茂木 康弘,原岡 たかし. (2013). Development of PSA technology for the separation of carbon dioxide from blast furnace gas. *JFE* 技報, *32*, 44.

铁诺咨询. 新日铁住金高性能CO_2回收装置 (ESCAP®) 的开发[EB/OL] (2014–10–17) [2015–08–20]. http://www.tncsteel.com/show.php?contentid=153907.

Part III
Market

Building Carbon Market in China: Take Stock and Look Ahead

5

Jun Fu, Xing Chen, and Jintao Xu

Addressing climate change and reducing CO_2 emissions has become a consensus globally. Among major industrialized countries, policy makers have either deployed or started to contemplate about deploying market-based instruments for climate policy. In that context, the idea of "cap & trade" of the carbon market—alternatively known as emission-trading system (ETS) in Europe—has commanded much attention of late. Theoretically, between the two extreme poles of hierarchy and market, their hybrid in the form of ETS can reduce CO_2 emission reductions at lower costs than direct administrative fiat such as mandated installation of emission-reduction technologies. Deployed singularly, the former is about quantities, whereas the latter is about prices.[1] But both are needed to fight against climate change if only because of the problem associated with free riding.

In line with the above, in 2011, China designated 7 provinces/cities, i.e., Beijing, Shanghai, Tianjin, Chongqing, Guangdong, Hubei and Shenzhen as ETS pilots (Duan et al., 2014). It was part and parcel of the country's strategy to

[1] For more discussion on this, see e.g., Weitzman (1974). Mideksa and Weitzman (2019).

The original version of this chapter was revised: Few words corrections on page 198 have been updated. The correction to this chapter can be found at https://doi.org/10.1007/978-981-16-4310-1_8.

J. Fu (✉) · J. Xu
National School of Development, Peking University, Beijing, China
E-Mail: fujun@pku.edu.cn

X. Chen
School of International Relations and Public Affairs, Fudan University, Shanghai, China

J. Fu et al. (eds.), *Climate Mitigation and Adaptation in China*,
https://doi.org/10.1007/978-981-16-4310-1_5

decarbonize its economy, and this move was the first phase (2011–2017) of a 3-phased plan to build carbon market in China. The vision was that these regional ETS pilots, if successful, would serve as stepping stones for China to prepare for a nation-wide ETS in the second phase (2017–2020), and following the second phase, China would consolidate and scale up the national ETS in the third phase (after 2020). Indeed, it was expected that China's ETS would eventually overtake EU-ETS to become the largest in the world, covering emission-trading of up to 6–9 billion tons of CO_2.[2] In China, whereas there is broad agreement among environmental economists about the advantages of market-based resolutions, there is much debate as to whether or not a unified national ETS is the best choice for the country in light of regional disparities in industrial structure across the country.

In any event, are these regional ETS pilots working, albeit still at an early and experimental stage? What lessons can we learn from these regional ETS for the benefits of designing and developing a national ETS in the years ahead?

This chapter aims to investigate these questions. The chapter is organized into four Sections. Section 1 starts with a brief review of the theoretical basis and the current practice of ETS worldwide and then outlines the evolving policy framework for China's ETS. Section 2 looks at the institutional designs and operational mechanics of these regional ETSs. In Section 3, adding rigor and going beyond a *prima facie* analysis, we deploy synthetic control method to evaluate the performance of China's regional ETSs. Key takeaways and policy recommendations follow in Section 4.

1 Section 1

1.1 "Cap & Trade": Theory and Practice of ETS

The theoretical basis of "cap & trade" of emission-trading system (ETS) can be traced back to the economic thinking of Nobel laureate Ronald Coase, which made it possible to frame the discourse on pollution control in terms of rights

[2] This is an estimation with the following assumptions: (1) China controls yearly CO_2 emissions under 10 billion tons in 2016–2020; (2) it peaks CO_2 emissions at 15 billion tons by 2030, and (3) the coverage of China's ETS will be at the same level of EU-ETS, i.e., 60%. Assuming that the turn-over rate is at 300%, the trading volume of China's ETS could reach 18–27 billion tons. In monetary terms, the value could reach 1.8–2.7 trillion RBM, assuming the price then is RMB 100/tCO_2. According to UNEP, China's CO_2 emissions were 9.839 billion tons in 2019, accounting for 27.2% of the world's total.

instead of taxes, as previously advocated by Arthur Pigou (Pigou, 1929). According to Coase's theme, by identifying certain rights, such as the right to emit greenhouse gas (GHG) emissions, and enabling the rights to be transferable, private negotiations could lead to the optimal allocation of resources, as long as transaction costs are sufficiently low (Coase, 1960).

Essentially, the "cap & trade" scheme affects levels of greenhouse gases (GHG) emissions by way of establishing a market for emission permits allocated to emitters under a total amount of emissions allowed—hence the "cap". The total amount of allowable emissions then is divided into some emission permits allocated to individual business entities under certain rules (Weitzman, 2012, 2017). Recognizing the marginal cost difference of different emitters, the scheme allows trading of emission permits to eventually arrive at an equilibrium price—hence the "trade". The equilibrium price provides a signal of the scarcity of emission permits (Spence and Weitzman, 1978), which in turn can guide individual emitters to choose between different strategies, e.g., increasing GHG emissions or reducing GHG emissions by investing in new technologies.

The first large-scale practical implementation of emission trading was the US sulfur dioxide (SO_2) "cap & trade" program. The program started in 1995 intending to address the problem of acid rain in the northeastern United States by mandating significant reductions in total annual SO_2 emissions from coal-fired power plants (Stavins, 2001). The success of the program resulted in the emission-trading scheme being adopted as a crucial element of the design of the Kyoto Protocol in 1997 (UNFCCC, 1998). Worldwide, the practice of "cap & trade" includes Chicago Climate Exchange (CCX), UK Emissions Trading Group (ETG), National Trust of Australia (NSW), and the EU GHG Emission Trading System (EU-ETS).

At present, the European Union (EU) is leading in terms of building GHG emission-trading systems. The EU-ETS started in 2005, and by 2015, it had covered 31 EU countries, accounting for 50% of EU's total carbon emissions (European Commission, 2015). As envisioned, from 2015 to 2020, the EU-ETS has been undergoing three phases of construction. During the first phase (2005–2007), the EU-ETS covers industries such as power, petrochemical, steel, construction materials, and paper-making, and it sets a yearly CO_2 emission cap at 2.299 billion tons. During the second phase (2008–2012), the EU-ETS expands its industry coverage to include aviation, and it sets a yearly CO_2 emission cap at 2.081 billion tons. During the third phase (2013–2020), the EU-ETS further expands its industry coverage to include chemicals and electrolytic aluminum, and it sets a yearly CO_2 emission cap at 1.846 billion tons.

Another salient feature of the EU-ETS is that it would gradually increase emission permits from free allowance to auction. Specifically, while auction does not exceed 5% in the first phase, it will increase to 10% during the second phase, and again to 30% during the third phase. The general idea was to increase the competitive force of the market overtime (European Commission, 2003).

1.2 Policy Framework for Regional ETS Pilots in China

Under the UN Framework Convention for Climate Change (UNFCCC), in 2009, China had pledged to reduce carbon intensity per unit of GDP by 40–45% in 2020 from the 2005 level. Subsequently, China incorporated the concept of "carbon intensity" into its 12th Five Year Plan (FYP, 2011–2015) for the first time, and the country decided to "*gradually* [emphasis added] establish a carbon emission trading market" to control GHG emissions. The wording with respect to ETS in the FYP indicated that China's approach would be a phased and gradual one. In other words, "testing it out and then scaling it up" was the way to go. In retrospect, China took its international pledge seriously, and the gradual approach for the construction of ETS was no surprise, as it followed the same logic of the overall reform and opening-up program of the whole country during the past 40 or so years.

In October 2011, the National Development and Reform Commission of China (NDRC) issued the "*Notice on Conducting Carbon Trading Pilots*"to seven provinces/cities in China—Beijing, Shanghai, Tianjin, Chongqing, Guangdong, Hubei and Shenzhen. In effect, this NDRC notice was an instruction for the seven provinces/cities to prepare for the institutional groundwork for the launch of ETS pilots in 2013–2014. The seven regional ETS pilots covered about 18% of China's population, and 30% of the country's GDP

Table 1 documents the legislative or policy framework at the seven localities to clear ways for the launch of the seven regional ETS pilots. As these regional ETS pilots were rolled out in 2013, it was expected that based on their experience, China would build a nationwide ETS to accelerate the process to decarbonize its economy. Indeed, according to the Joint China-US Presidential Statement in 2015, China promised to launch the national carbon market by the end of 2017.

Table 1 Policy framework of regional ETS pilots in China

Localities	Key documents
Beijing	Resolution of Beijing Municipality to Carry out Carbon Trade Pilot and to Control Carbon Emissions (Standing Committee, Beijing Municipal People's Congress)
Shanghai	Shanghai Carbon Emissions Management Provisional Guidelines (Shanghai Municipal People's Government, November 18, 2013)
Guangdong	Guangdong Carbon Emissions Management Provisional Guidelines (Guangdong People's Government, January 15, 2014)
Shenzhen	Regulation of Shenzhen Special Economic Zone on Carbon Emissions Management (Shenzhen Municipal People's Congress, December 30, 2012)
Tianjin	Interim Measures of Tianjin Carbon Emission Trading (General Office of Tianjin Municipal People's Government)
Hubei	Hubei Interim Measures on Carbon Emissions Trade Management (Hubei Provincial Government, April 23, 2014)
Chongqing	Chongqing Carbon Emissions Trade Management Interim Measures (Chongqing Municipal People's Government, March 27, 2014)

Source Websites of local DRC of Beijing, Shanghai, Tianjin, Chongqing, Guangdong, Hubei and Shenzhen

2 Section 2

2.1 Institutional Designs of ETS Pilots

Good institutional designs were critical for the success of the seven regional ETS pilots in China, and eventually, for the success of the national ETS, which China planned to launch by the end of 2017. As discussed in the previous section, China embarked on building carbon markets more or less from scratch, as a long-term strategic move to develop a low-carbon economy in view of the country's pledge in 2009 to the international community to reduce carbon intensity of GDP by 40–45% by 2020 from the 2005 level. After a couple of years of initial preparations starting in 2011, China started seven regional ETS pilots in 2013.

These regional ETS pilots were intended for the country to learn from experience before moving on to build a nationwide ETS. Given regional disparities in industrial structure, the advantage of such a strategy—i.e., regional pilots rather than one-swoop-go—is that local governments had the flexibility to design, operate, and adjust their ETS in light of local conditions. In Chinese aphorism, it is "crossing the river by touching the stones underneath", so to speak, and there was a lot of "learning by doing" and "learning by watching" going on in the initial years. Indeed, during the initial phase, the regional ETS pilots had all learnt to expand industry coverage, improve permit allocations, and develop derivative products. And as a reflection of their demonstration effects, other provinces (e.g., Jiangsu, Xinjiang, and Zhejiang) increasingly showed interest in having their own ETSs. Fujian province, for instance, launched its own ETS in late 2016.

A brief observation shows that China's regional ETS pilots exhibit the following institutional features:

1. These regional ETS pilots fall under the local Development and Reform Commission's (DRC) jurisdiction, and the Local Finance Bureaus.
2. All regional ETS pilots generally distribute permits in a fashion of grandfathering (free distribution), except Guangdong, Shenzhen, and Hubei where permits are auctioned.[3]
3. The government regularly intervenes in these pilot markets for data collection, allocation of permits and/or auction, and pricing if need be.
4. The regional ETS pilots cover mostly energy-intensive industries such as electricity, steel, cement, chemicals. ETSs in Beijing, Shanghai, and Tianjin, however, also cover building and service sectors. In all, some 20 industries are covered.[4]
5. All ETS pilots only cover CO_2 emissions, except Chongqing, which covers 6 GHGs (CO_2, CH_4, N_2O, HFCs, HFCs, SF_6). Most ETS pilots cover both direct and indirect emissions, although indirect emission definition differs. Some account for purchased electricity only; others cover purchased electricity *and* heat. Hubei only trades direct emissions.
6. All regional ETS pilots recognize China Certified Emission Reduction (CCER) by NDRC, although some set ceilings for their use, ranging from 50 to 70%.
7. All regional ETS pilots set up two thresholds for market access: transaction threshold (for emission control firms) or reporting threshold (for reporting firms). The latter, albeit not included in trading yet, must report emissions to the government to consider future inclusion (Fig. 1).

[3] Shanghai ETS represented another alternative, i.e., grandfathering + auction. See compliance/sanction below.

[4] By 2020, close to 3,000 fossil-heavy industrial firms covered.

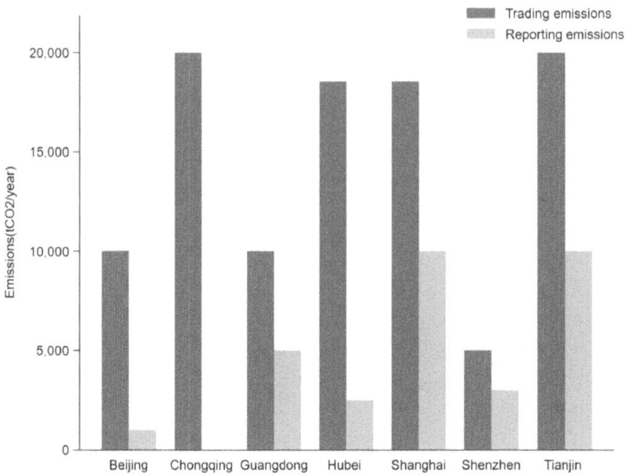

Fig. 1 ETS Pilots in China: Transaction Threshold and Reporting Threshold. *Note* 1. Data. *Source* Websites of local DRC in 7 regional ETS pilots; 2. Reporting emissions of Chongqing is unavailable

2.2 ETS Pilots in Operation

Institutional designs aside, let us take a look at the actual performance of China's regional ETS pilots. For the purposes of this study, we look at the five domains as key performance indicators (KPI) of the ETS pilots: (1) emission permit allocation, (2) MRV (monitoring, reporting and verification), (3) compliance/sanctions, (4) CCER (China certified emission reduction), (5) behavioral change, and (6) pricing. Table 2 lists the six categories of KPIs. Below, we schematically elaborate a bit more on each of the six categories of KPIs.

Table 2 China's regional ETS pilots in operation

Tasks				
Allocation of allowances	Indicator	Control agency	Data source	Status check
Time & allocation of permits to firms	Total amount of permits allocated	Local DRC	Local DRC website	Permits allocated in 2013 and 2014, with 730 million tons and 1285 million tons respectively
Information collection	Indicator	Control agency	Data source	Status
Collection of emission reports including verification	Total number of emission reports collected	Local DRC	Local DRC Website; interview with ETS staff	Collected emission and verification reports from 1736 firms in 5 ETS pilots[a]
Random checks on verification reports	Total number of random checks	Local DRC	Local DRC Website; interviews with ETS staff	No data on random checks on verification reports
Information disclosure on trading volume and price	Publication patterns of trading volume and price	Local DRC or carbon trading exchange	Local DRC Website; trading exchange; K-line website	All carbon trading exchanges released all information on trading volume and price except Guangdong, whose data is only on most recent trading

(continued)

Table 2 (continued)

Tasks				
Allocation of allowances	Indicator	Control agency	Data source	Status check
Information disclosure on compliance	Information compliance rate	Local DRC	Local DRC Website	All ETS pilots published information on compliance within 3 months after a compliance cycle
Compliance and sanction	Indicator	Control agency	Data source	Status check
Rules and sanctions on non-compliance	Total number of sanctions imposed	Local DRC	Local DRC Website; news reports	14 firms were sanctioned for non-compliance. Nature of sanction unknown
Other policy instruments	Indicator	Control agency	Data Source	Status check
China Certified Emission Reduction (CCER)	Total CCER issued	NDRC	China CCER Exchange Info-Platform	A total of 13.72 million tons of CCER were issued by Jan. 14, 2015
Behavioral change				
Firms behavior	Indicator	Control agency	Data source	Status check

(continued)

Table 2 (continued)

Tasks				
Allocation of allowances	Indicator	Control agency	Data source	Status check
Behavioral changes or adaptions	Amount of emission permit traded	Local Carbon Trading Exchanges	Local Carbon Trading Exchanges	A total of 114.58 million tons of permits was traded by June 30, 2017
Adaption to compliance	Compliance rate	Local DRC	Local DRC Website	Compliance rates in 2013: Beijing 97%, Shanghai 100%, Guangdong 98%, Shenzhen 99%; Tianjin 96%

[a]415 firms in Beijing, 191 in Shanghai, 184 in Guangdong, 114 in Tianjin, and 635 firms and 197 buildings in Shenzhen
Source Websites of local DRC in 7 regional ETS pilots

1. Emission Permit Distribution

CO_2 emission permit allocation is a prerequisite for the carbon market to operate. Once the government sets the cap of emissions, it has to find a way to allocate allowances to emitters in a timely fashion (Pang and Duan 2016). The way that the regional ETS pilots in China operate in this area was that the government of each pilot launched a dedicated e-platform or website for registration and allocation of emission permits. By government, it referred to local DRC in each case. Allocation of emission permits had two ways: grandfathering and auction. In the case of Guangdong, Shenzhen, and Hubei, it was auction; and in the case of Beijing, Chongqing, Shanghai, and Tianjin, it was free distribution. In all, free distribution accounted for over 90%. There was room for the websites of these regional ETS pilots to improve. A quick browse in a comparative perspective indicated that these e-platforms were not equally user-friendly. For instance, the website of Beijing did not disclose data on how permits were allocated in 2013 and 2014.

2. MRV System

Transparency and accuracy of information is essential for market transactions. In the case of ETS, a robust MRV (monitoring, reporting and verification) system for corporate ESG (environment, social responsibility and corporate governance) is vital to minimize the risks of "greenwashing". Here, it is the role of the government to collect information at the firm level with respect to emissions, trading, compliance, etc. Currently, all ETS pilots in China would require emission control firms to submit emission reports verified by a third party designated by the government. These reports then would be subject to random checks by the government to increase accuracy and reliability. Again, a quick browse of the websites of these regional ETS pilots showed that whereas information on trade volume and price was readily available, that of compliance was patchy, and that of verification was not available—a critical area that need further investigation.

3. Compliance/Sanctions

A robust regulatory framework has to put in place the so-called "teeth" of enforcement so that it can encourage or rather compel firms to take emission limits seriously and participate in carbon trading actively. Here, the rules regarding compliance and sanctions across these regional ETS pilots were vague, although there were sporadic reports on sanctions. For instance, in late 2013, Beijing and Guangdong imposed sanctions on a total of 14 firms for their failure of timely compliance, although little information was available as to what kind of penalty was imposed. In Tianjin, there was report that 4 firms had failed to perform their

2013 obligations in a timely fashion, and yet Tianjin Interim Measures did not contain penalty clauses, fines or otherwise. In Shenzhen, there were reports that grace periods were granted by the government to urge for compliance. It seemed that in terms of compliance and sanctions, Beijing and Guangdong were the most stringent, whereas Tianjin was most lax in enforcement of rules.

Shanghai ETS represented another alternative. Apparently, there was no case of non-compliance or sanctions in 2013. But the way that Shanghai ETS operated was that when the deadline (June 30) for compliance was approaching, the government auctioned off additional 580,000 tons of CO_2 emissions permits that firms could buy for compliance. The result was high compliance rate.

4. CCER (China certified emission reduction)

All seven regional ETS pilots recognized project-based CCER (China certified emission reduction) approved and issued by NDRC, and information for authenticity and verification of these credits was available at China CCER Exchange Platform. In 2016, in terms of project number, Hubei ETS had 122 projects, Beijing ETS had 16, Chongqing ETS had 15, Guangdong ETS had 47, Shanghai ETS had 18, Tianjin ETS had 6, and Shenzhen ETS had 5, with a total of 229. Prominent among these projects were wind, solar, hydraulic power, and methane gas.[5] While the set ceiling for the use of CCER credits varies across the seven regional ETS pilots, ranging from 1 to 10%,[6] CCER credits could count as emission allowances for compliance.

5. Behavioral Change

Does ETS change firms' behavior, and if yes, to what extent? Are there variations across the seven regional ETS pilots? For the purposes of this study, we look at three key indicators—trading volume, trading turnover, and active ratio—defined as the ratio between the number of days that have trading volume and the total trading days. Our statistical deadline is June 30, 2017 and the results are tabulated in Table 3.

[5] They accounted for over 80% in terms of project number in 2020.

[6] Specifically, offset limit of CCERs as a share of annual allowance was set at 10% for Hubei, Shenzhen, Tianjin, 8% for Chongqing, 5% for Beijing, and 1% for Shanghai. Note also that credits from hydro projects are not allowed in most ETS pilots. In the case of Guangdong, the ceiling was set at 1.5 million offsets, and both CCERs and Pu Hui Certified Emission Reduction (PHCERs) were accepted, but at least 70% had to come from local projects in Guangdong province.

Table 3 ETS Pilots: trading indicators

ETS Pilots	Starting date	Total trading days	Active ratio (%)	Average price(rmb)	Average trading volume(ton)	Average trading turnover(rmb)	Share of total (%)
Shenzhen	Jun-13	1885	89	38	14,349	546,882	12
Beijing	Nov-13	1765	64	60	8169	488,521	6
Shanghai	Dec-13	1750	62	31	10,183	315,377	8
Guangdong	Dec-13	1750	86	24	43,495	1,023,164	32
Tianjin	Dec-13	1744	38	18	5827	105,497	4
Hubei	Apr-14	1674	96	23	47,771	1,101,884	34
Chongqing	Jun-14	1623	35	17	5392	92,240	4
Average		1742	67	30	19,312	524,795	–

Note 1. Active rate means the ratio between number of days that have trading volume and total trading days; 2. The statistical deadline is set at Dec 31, 2020; 3. Total trading days excludes holidays; 3. *Data source* http://k.tanjiaoyi.com/..

The Table 3 shows that in terms of trading volume, Hubei was the highest and Tianjin was the lowest; in trading turnover, Hubei was the highest and Tianjin was the lowest; and in terms of active ratio, Hubei, again, was the highest and Tianjin was the lowest. Across the seven regional ETS pilots, the average active ratio was 66%. Above that ratio, there were Hubei (96%), Shenzhen (90%), Guangdong (71%), and Beijing (69%), and below that ratio, there were Shanghai (63%), Tianjin (52%), and Chongqing (18%).

It seemed clear that, having operated for more than 3 years, the 3 ETS pilots, i.e., Hubei, Guangdong, and Shenzhen, were more active in trading than the other 4, i.e., Shanghai, Beijing, Chongqing, and Tianjin. Amongst the seven regional ETS pilots, Hubei ETS, in particular, was relatively successful, in terms of trading volume and turnover. By June 30, 2017, Hubei ETS sold 40.48 million tons of permits, with a turnover of RMB 797.82 million. But by international standards, even the best of China's ETS pilots still had a long way to go. For instance, on a single day on January 21, 2015, the trading volume of California carbon market surged to 10 million tons, about of 2/3 of China's total trading volume in all seven regional pilots (Fig. 2).[7] In the case of China, as Fig. 2 shows, the sudden and one-time spikes in trading volume around June 2015 indicate that many firms perhaps

[7] http://calcarbondash.org/.

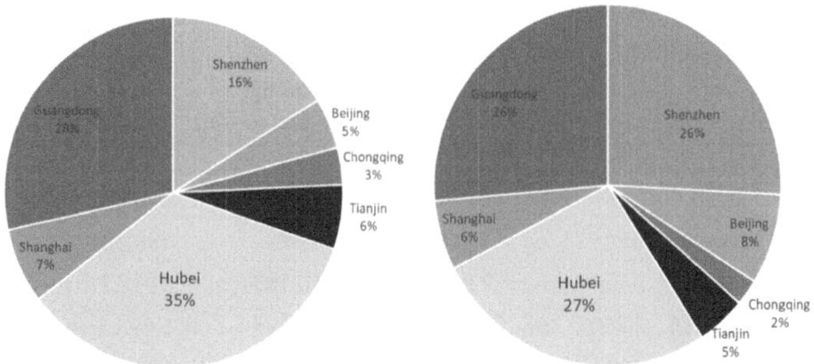

Fig. 2 Share of trading volume/trading turnover. *Data source* http://k.tanjiaoyi.com/

take emissions trading as political cooperation rather than investment opportunity. Carbon assets were not considered valuable.

6. Pricing

Price is central to market (also see Fig. 2), and yet, given the fact that the regional pilots of carbon market in China remained divided, it was expected that carbon prices would vary across different regions. Indeed, there was a great variation of carbon price among these ETS pilots with a minimum of RMB 28 and a maximum of RMB 143 per ton of CO_2 emissions. By June 30, 2017, the total trading volume of the seven ETS pilots reached 114.58 million tons. As Table 2 shows that amongst the seven regional ETS pilots, average price for CO_2 emissions was the highest in Beijing, about RMB 50/ton, followed by Shenzhen, with RMB 47/ton. The price in Shanghai, Chongqing, Hubei, and Tianjin was close, about RMB 20–25/ton. As an average of all seven ETS pilots, it was about RMB 30/ton—a price perhaps too low (still trending downwards, see Fig. 2) to significantly motivate a firm's strategic investment in shifting towards better technology or cleaner energy. To highlight the point here, the current cost of CCUS technology is estimated at US$ 600/ton.[8] Moreover, the price divergence across different regions was also indication of the great challenges lying ahead in setting up a unified carbon market in China.

[8] For more discussions on carbon capture, utilization and storage, see Chapters 3 and 4 of this study.

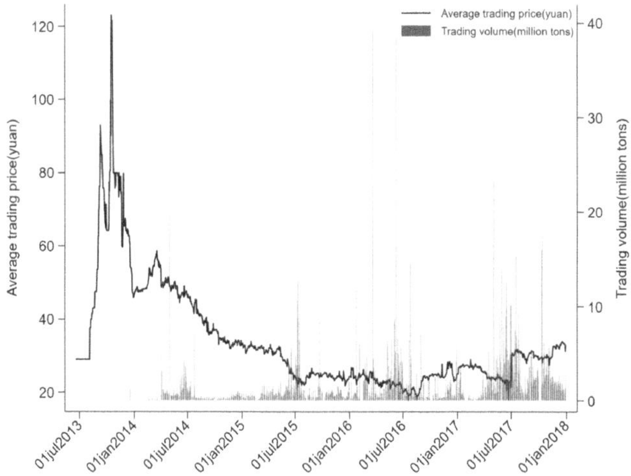

Fig. 3 Trends of emission permit trading in China. *Data source* http://k.tanjiaoyi.com/

For international comparative perspectives, average carbon price in EU-ETS was about RMB 100/ton (or US$ 16/ton) before 2018 and poised to rise significantly;[9] globally, the price range was between US$ 1/ton and US$ 130/ton (Fig. 3).

3 Section 3

3.1 Assessing CO_2 Emission Reduction by ETS Pilots

In the previous two sections, we have reviewed the theoretical basis of carbon market and looked at the institutional designs and operational mechanics of China's regional ETS pilots. In this section, building upon that knowledge, our aim is to go beyond mere descriptive statistics to assess the actual performance of the ETS pilots in reducing CO_2 emissions. To increase the rigor of our examination beyond a *prima facie* case, we will deploy the method of synthetic control in our statistical analyses, trying to pin down the impact, if any, of the regional

[9] It reached 25 Euro per ton in 2019.

ETS pilots on the reduction of CO_2 emissions. Strictly speaking, doing scientific experiments means that one must isolate variables, and in so doing, tests must proceed under strictly (often lab-) controlled conditions. The same logic would apply to serious policy studies by a trained eye—he or she must be keenly aware that real world knowledge based on inductive inference can never be deterministic, and worse, "controlled conditions" in policy settings at the aggregate level are elusive at best if not totally absent.

Mindful of these pitfalls, in the three subsections that follow, we will first introduce and then deploy the method of synthetic control. Second, we report investigation results based on that method. Third, we perform a placebo-like robustness test to check the confidence level of whatever impact we may find of the ETS pilots on the reduction of CO_2 emissions.

3.2 Method of Synthetic Control

Frequently researchers are called upon to examine the effects of policy interventions, and in so doing they often resort to comparative case studies. But that job is easier said than done. For the challenge is that at the aggregate level how can the researcher seriously compare the results of an allegedly "affected" group vis-à-vis an allegedly "unaffected" group, when in fact there is no controlled group? Indeed, in our present case of ETS and CO_2 emissions, it would be difficult to estimate CO_2 reduction achievement of carbon trading in the absence of solid control groups.

In this context, we need relevant statistical tools and the method of synthetic control is helpful in estimating effects of policy settings where a single group (or a firm or a state) is exposed to policy intervention. The method of synthetic control was first developed and used by Abadie et al. Studies, employing the synthetic method, include the assessment of the economic impact of conflict in Basque (Abadie and Gardeazabal, 2003), the impact of tobacco control program on tobacco sales in California, USA, (Abadie et al., 2010), the economic impact of the 1990 German reunification on West Germany (Abadie et al., 2015). There are also many studies on China. Deploying synthetic control method, Liu and Fan (2018) have examined the economic impact of property tax in Chongqing. Zhang et al. (2016) have tried to respond to the query—"Did Olympic Games improve air quality in Beijing?" by using synthetic control method.

Drawing upon Abadie et al. (2010), for the purposes of our investigation of ETS, we build a model that has the rationale for synthetic control method. Assume that we have $M+1$ regions, and region # 1 is exposed to policy intervention (i.e., ETS in our case), then we have the remaining M regions as control units. Let T_0 be the number of pre-intervention periods, with$1 < T_0 < T$. Assume Y_{it}^N is the counterfactual outcome that would be observed for region i at time t with the absence of policy intervention for units$_i$, and time periods$_t$. Let Y_{it}^I be the outcome that would be observed for region i at time t with policy intervention in afterT_0. We also assume that the policy intervention has no effect on the outcome before the implementation period, so that for all i in period$t \in \{1, 2, \ldots, T_0\}$, we would have that $Y_{it}^N = Y_{it}^I$. Let $\alpha_{it} = Y_{it}^I - Y_{it}^N$ be the effect of the policy intervention for unit i at time t. Our aim here is to estimate:

$$\alpha_{1t} = Y_{1t}^I - Y_{1t}^N$$

where Y_{it}^I is observed, so we only need to estimateY_{1t}^N. Let (M×1) be a vector of weights $W^* = (w_2^*, \cdots, w_{j+1}^*)$ such that $W_m \geq 0$ for m $= 2, \ldots,$ M + 1 and$w_2 + \cdots + w_{M+1} = 1$. Each value of the vector W^* denotes a potential synthetic control, which corresponds to a particular weighted average of control regions. The value of the outcome variable for each synthetic control indexed by W^* is:

$$\sum_{m=2}^{M+1} w_m Y_{it} = \partial_t + \theta_t \sum_{m=2}^{M+1} w_m Z_m + \lambda_t \sum_{m=2}^{M+1} w_m \mu_i + \sum_{m=2}^{M+1} w_m \varepsilon_{it}$$

Abadie and his associates (2011) have proved that $\sum_{j=2}^{J+1} w_j^* Y_{jt}$ is an unbiased estimate of Y_{it}^N, where $T_0 < t \leq T$. So $\alpha_{1t} = Y_{it} - \sum_{j=2}^{J+1} w_j^* Y_{jt}$ is the unbiased estimation of α_{1t}.

For the purposes of this investigation, the outcome variable of our interest is annual CO_2 emissions. We calculated CO_2 emissions based on energy consumption at provincial level by using energy consumption data in 1997–2017 period (Provincial Statistical Yearbook). Since China's seven regional ETS pilots began operation in 2013, we mark 2013 as the year of policy intervention (i.e., ETS).

This gives us 16 years of pre-policy intervention data. Based on the method proposed by CEADs,[10] we calculate annual CO_2 emission level for all provinces in China.

To see the "synthetic" part of the method as a benchmark devise, let us take Hubei as an example. We use the synthetic control techniques described above to construct a "synthetic Hubei". This synthetic Hubei is a best weighted average with a set of value predictors of carbon emissions in Hubei before policy intervention (i.e., ETS) began. In this context, the "synthetic Hubei" is meant to indicate the level of carbon emissions that we would have observed in the absence of ETS. We thus drop from the donor pool all the provinces/cities that have adopted ETS during the sample period (i.e., Beijing, Guangdong (including Shenzhen), Shanghai, Tianjin, and Chongqing). However, since some key economic indicators of Tibet and Ningxia are not available during sample period, a total of 23 provinces/cities have remained in our donor pool (Table 5).

Using the techniques described above, we construct a synthetic province/city one by one for Hubei, Guangdong (including Shenzhen), Tianjin, Shanghai, Beijing, Chongqing that mirrors the level of carbon emissions before ETS was put in place. Based on that, we estimate the emission reduction effect of policy intervention of ETS after 2013 as the difference in carbon emissions between real Hubei (with ETS) and synthetic Hubei (without ETS). We then perform a series of placebo tests to see whether or not the estimated effects of ETS are unusually large in light of the estimate distribution that we obtain by applying the same procedure to the provinces/cities in our donor pool. Finally, we repeat the same procedure for other ETS pilots—Guangdong (including Shenzhen), Beijing, Shanghai, Chongqing, and Tianjin.

3.3 Test Results

As explained above, by deploying the synthetic control method, we construct a synthetic Hubei as the combination of provinces/cities in our donor pool. This synthetic Hubei resembles Hubei in terms of predictor of CO_2 emissions before policy intervention (i.e., ETS). Table 4 exhibits the test results, which have compared the features of real Hubei with synthetic Hubei before policy intervention. The same comparison includes the other regional ETS pilots. Table 5 shows the weights of all control provinces/cities for each ETS pilot.

---

[10] China Emission Accounts & Datasets: https://www.ceads.net/. CEADs is made of a group of experts from the UK, the US, and China who have worked on China's emission accounting

Table 4 Economic features of real pilots versuss synthetic pilots

	HB	HB_syn	BJ	BJ_syn	SH	SH_syn	TJ	TJ	CQ	CQ_syn	GD	GD_syn
GDP	7862.2	7887.3	7172.3	5186.2	9186.0	9141.2	4385.7	3414.2	4029.2	4048.1	23,489.4	17,520.6
GDP_per	13,703.1	13,746.8	42,862.0	17,454.5	47,094.2	21,964.8	37,534.3	14,944.7	13,678.4	13,286.3	24,596.6	24,505.5
Population	5742.4	5759.0	1544.7	2108.6	1848.4	3309.1	1086.8	1979.9	2850.1	2860.6	9033.3	6532.9
Sec_rate	42.6	42.8	29.9	33.5	46.5	50.4	52.8	48.9	44.2	44.3	48.4	49.0

note 1 sec rate refers to second industry as share of total GDP; 2. *Data source* China statistical yearbook

Table 5 Synthetic weight for synthetic pilots

Region	Hubei_weight	Beijing_weight	Shanghai_weight	Tianjin_weight	Chongqing_weight	Guangdong_weight
Anhui	0.036	0	0	0	0.027	0
Fujian	0.025	0	0	0	0.03	0
Gansu	0.019	0	0	0	0.047	0
Guangxi	0.033	0	0	0	0.031	0
Guizhou	0.025	0	0	0	0.042	0
Hainan	0.096	0.682	0	0	0.08	0.144
Hebei	0.026	0	0	0	0.02	0
Heilongjiang	0.018	0	0	0	0.034	0
Henan	0.045	0	0	0	0.015	0
Hunan	0.059	0	0	0	0.023	0
InnerMongolia	0.024	0	0.018	0	0.036	0
Jiangsu	0.057	0	0	0	0.012	0.856
Jiangxi	0.025	0	0	0	0.034	0
Jilin	0.017	0	0	0.566	0.04	0
Liaoning	0.024	0	0	0	0.026	0
Qinghai	0.013	0	0.352	0.381	0.301	0
Shaanxi	0.02	0	0	0	0.036	0
Shandong	0.027	0	0	0	0.01	0
Shanxi	0.017	0	0	0	0.039	0

(continued)

Table 5 (continued)

Region	Hubei_weight	Beijing_weight	Shanghai_weight	Tianjin_weight	Chongqing_weight	Guangdong_weight
Sichuan	0.322	0	0	0	0.02	0
Xinjiang	0.021	0	0	0	0.043	0
Yunnan	0.025	0	0	0	0.036	0
Zhejiang	0.029	0.318	0.629	0.052	0.02	0

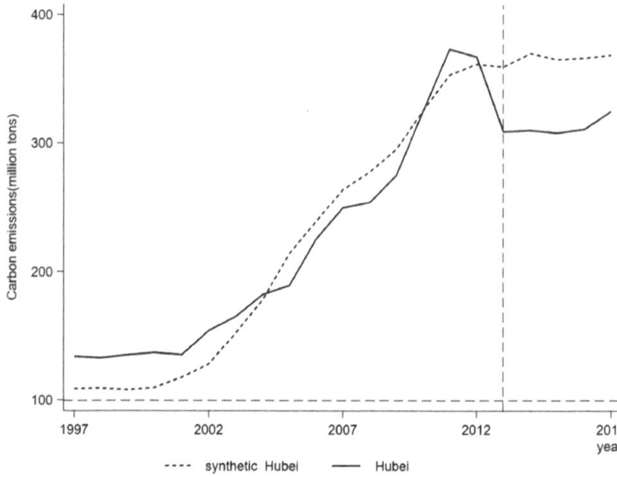

Fig. 4 CO_2 emissions between Hubei and synthetic Hubei

Due to significant variations of social and economic conditions across different regions in China, we have divergent synthetic results. Generally speaking, the closer to the China's average level, the better the synthetic results. For instance, Hubei and Guangdong have displayed better synthetic results than that of the other pilots.

Figure 4 shows the trend of carbon emissions in real Hubei and synthetic Hubei. The result shows that after 2013, carbon emissions in real Hubei and synthetic Hubei diverged significantly. For impact assessment at a higher level of granularity, so that causal link has a higher degree of certainty, we also apply the same procedure to specific industries, i.e., non-ferrous metals, ferrous metals, electricity industry and six polluting industries[11] in Hubei ETS pilot. The results also demonstrate that these industries in Hubei have all reduced carbon emissions substantially (Fig. 5).[12]

methods and applications. They provide most up-to-date carbon emission, social, economic, and trade data.

[11] Six polluting industries include electric power, ferrous metals, nonferrous metals, raw chemical materials and chemical products, nonmetal mineral products, petroleum processing and coking.

[12] High granularity, industry-specific investigations are particularly important because they may yield insights in terms of cross-region homogeneity or heterogeneity. These insights, if

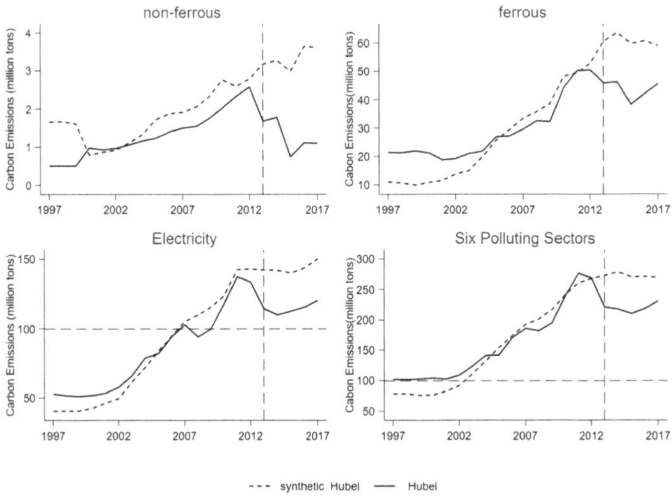

Fig. 5 CO_2 emissions of polluting industries between Hubei and synthetic Hubei

Likewise, Fig. 6 shows that before 2013, the synthetic data fits the actual data reasonably well in the case of Guangdong as well. After 2013, however, the two lines began to diverge significantly, indicating that Guangdong ETS pilot had made a significant difference in the reduction of carbon emissions.

In the previous section, we have learnt that in terms of trading volume, trading turnover, and active ratio, Hubei and Guangdong (including Shenzhen) scored higher than other regional ETS pilots. Our synthetic test results support our expectation or validate our hypothesis, that is, more trading volume leads to higher levels of emission reduction.

By contrast, the case of Tianjin shows very different results. Figure 7 plots trends of carbons emissions in real Tianjin and synthetic Tianjin. However, the test result shows that since 2013, carbon emissions in real Tianjin and synthetic Tianjin did not differ notably. The result comes as no surprise. As discussed in the previous section, Tianjin ETS pilot was way below average among the seven regional ETS pilots in many performance indicators.

When the same procedure is applied to Beijing, Shanghai, and Chongqing, we cannot get clear-cut simulation results, indicating that it is perhaps too early

---

any, can help in designing strategies of convergence between regional and national ETSs in the future.

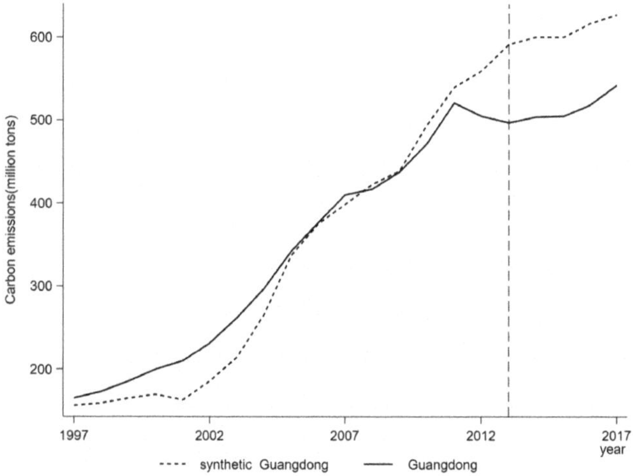

Fig. 6 CO_2 emissions between Guangdong and synthetic Guangdong

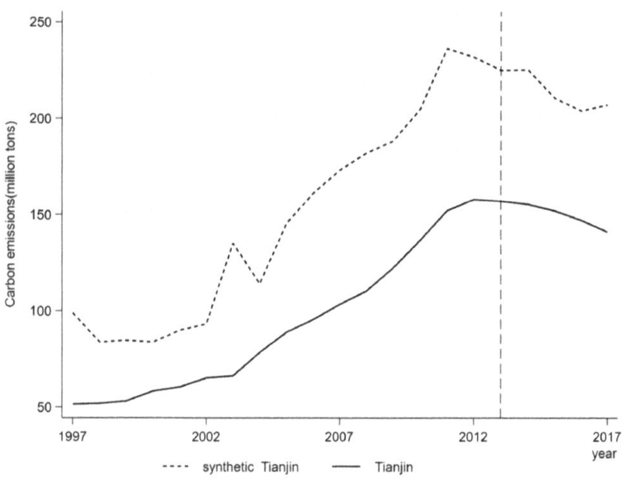

Fig. 7 CO_2 Emissions between Tianjin and synthetic Tianjin

to assess the actual performance of their ETS pilots at this moment. Or perhaps there are other "noises" that are not factored in and controlled for in our tests. For instance, in the case of Beijing, the dramatic bifurcation between real Beijing and synthetic Beijing had stated around 2008, indicating that there had been much more powerful forces going on other than mere the setup of ETS (Fig. 8). One plausible explanation may have to do with the hosting of Olympic Games in 2008, which had prompted Beijing to relocate much of its manufacturing out of Beijing proper.

And in the case Shanghai, the bifurcation—also very dramatic—had happened even earlier, around 2002 (Fig. 9), which warrants further investigation. Anyway, we must continue to watch Shanghai ETS carefully. Shanghai's sophisticated dual deployment of grandfathering and auction of CO_2 emission permits and its high compliance rate indicates that it has a nuanced understanding the role of state vis-à-vis that of the market, which normally would lend the city well to address issues related to climate change. After all, it takes both the state and the market to deal with the challenge of climate change and, as our theoretical discussion in Section I indicates, at the axiomatic level of carbon market is politics or public policy. It ultimately goes back to the question raised by Ronald Coase, i.e., where do we draw the line between the state and the market?

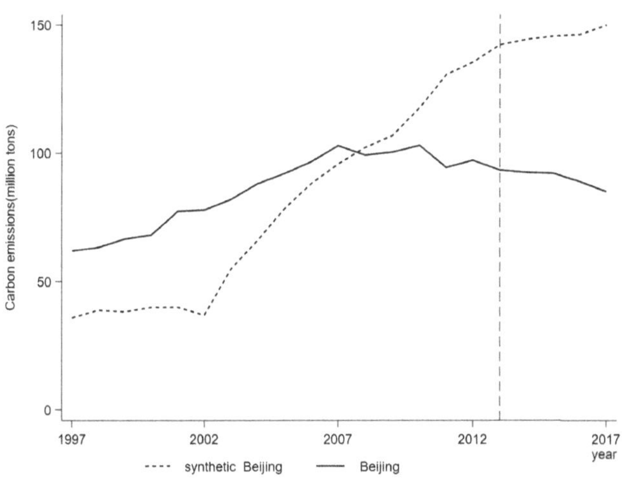

Fig. 8 CO_2 emissions between Beijing and synthetic Beijing

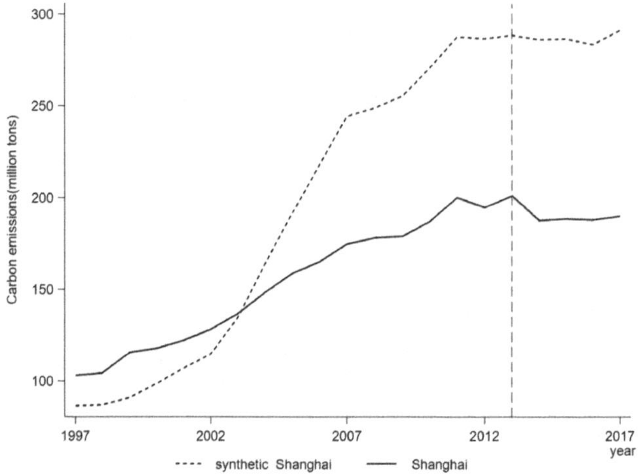

Fig. 9 CO_2 emissions between Shanghai and synthetic Shanghai

By comparison, the test result for Chongqing is not compelling, and we put it in the Appendix.

3.4 Robustness Test

To increase confidence level to our analyses, we also perform robustness test here. Let us begin by looking at Hubei ETS pilot again. As previously discussed, the test results we obtained by deploying the synthetic control method demonstrated that there was a significant gap of carbon emissions between synthetic Hubei and real Hubei. But how reliable are these results? As an exercise of inductive inference, our knowledge thus obtained can never be deterministic. Indeed, it all boils down to a matter of probability. In this context, a placebo test is appropriate, and it can be regarded as a permutation test in statistics. Our hypothesis testing here is to check statistical significance of our results. In other words, what is the probability that other provinces/cities showing as large a gap as Hubei ETS pilot does between Hubei and synthetic Hubei?

To answer the question, we employ the synthetic control method to estimate the effect of policy intervention of ETS on every other province/city. First, we need to drop the provinces/cities that show a poor fit for original data prior to

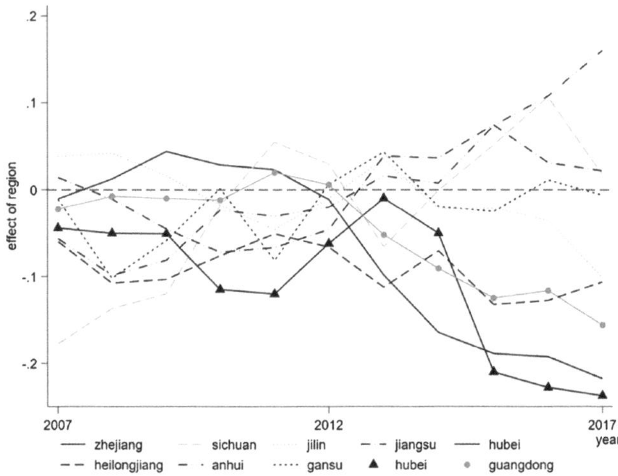

Fig. 10 Placebo test

the year when ETS began. Our concern is that even if the gap may not be caused after the policy intervention, it may result from the poor fitness before the policy intervention. Therefore, we have to drop the provinces/cities whose MSPE (mean square percent error) prior to 2013 is larger than 100. Finally, having done these procedures, we have 22 provinces in total as potential control units.

Figure 10 displays the placebo or permutation test results.

Here, each province/city is taken as a policy-intervened entity each time, with systematic comparison of carbon emissions between real and synthetic situations. From the test results, we can see that Hubei province is below the zero lines, which indicates that the probability of seeing a gap of Hubei's level in our dataset is 1/22, or statistically significant only at 4.5% or so. Conversely, these test results would increase our confidence level when we say that policy intervention in Hubei in the form of ETS has made a difference in reducing carbon emissions, even though it is still in early stage.

However, the same thing cannot be said—i.e., with the same level of confidence—for Chongqing (see Fig. 11 in Appendix)and Tianjin at this stage of development. In the case of Beijing and Shanghai, as discussed above, their situations seem to be more complex, and further investigation (e.g., field research) is needed. For the time being, however, in the case of Shanghai, the mixed use of grandfathering and auction in terms of carbon emission permits, together with high rates of compliance, indicates that Shanghai, itself already a financial center, is more sophisticated in setting up a robust MRV system in a cost-effective way. The point here is that while MRV system is vital to ETS, MRV itself is not transaction cost-free. Thus, efforts should be made to minimize the transaction cost in the future design of China's national ETS.

4 Section 4

4.1 Main Takeaways

At this early stage of construction of carbon market in China, there is initial evidence that the regional ETS pilots are not liquid enough to function well. Trading was low, price was low, and firms did not appear to be enthusiastic about carbon trading. For one thing, information disclosure and transparency have much room for improvement. For another, as regional ETS pilots, policy constraints exist and indeed prohibit cross-region emissions trading, and as a result, carbon price varies significantly across different regions. Arguably, this has made China's carbon market less attractive to potential investors, especially large institutional investors.

There is also evidence that the level of awareness of and participation in these regional ETS pilots has remained low. It is especially noteworthy that in the month just before the compliance date there is a sudden and one-time high trading volume. This perhaps indicates that most of the firms would take emissions trading as political cooperation rather than investment opportunity. Carbon assets are not considered valuable for many firms, which arguably is both a cause and effect of low carbon price in China. Overall, the average carbon price among all seven ETS pilots remained low, about RMB 30/ton—perhaps too low for firms seriously to consider strategic shifts towards a low-carbon growth path.[13] By comparison, carbon price in the US and EU was about US$ 15–16/ton (World Bank Group, 2018) and is expected to rise rapidly in the years to come.

[13] However, in 2020 the Chinese government made a new pledge to international community to peak carbon emission prior to 2030 and achieve carbon neutrality by 2060. In light of this, it is expected that carbon price would rise more rapidly under much harder constraints of "cap and trade" scheduled over the next 40 years, thus generating more pressure for fossil-fuel heavy firms to plan and shift more aggressively towards green growth. The new pledge also indicates that the centrality of China's climate policy in terms of international commitments has shifted from one that emphasized "carbon equity" to one that aims at "carbon neutrality".

Our analyses show that while success of these regional ETS pilots at this stage is very limited, it still varies considerably across different regions. Among the seven regional ETS pilots, Hubei is the best performer, and it has reduced about 59 million tons of CO_2 emissions in 2015. Guangdong and Shenzhen have done reasonably well, each having reduced about 37 million tons of CO_2 emissions in 2015. In the case of Beijing and Shanghai, the effects of ETSs, while not immediately visible, may have been overwhelmed by earlier policy inventions (e.g., holding 2008 Olympics in Beijing), which warrants further (field-trip) investigations beyond our current setup of econometrics. However, Tianjin has performed poorly during the same period.

Why? Some preliminary explanations are handy. Hubei has been a province with a progressive environmental agenda. In 2009, a pilot program for converting punitive levy of pollution to environmental tax was launched in Hubei. Zheng's research (2015) indicates that such pilot programs significantly reduced domestic pollutions, which might also have better prepared industrial firms in Hubei for the carbon trading scheme. During the whole era of reform and opening-up, Shenzhen has always been China's leading city in innovations and was China's first city to realize its "Blue Sky Plan." The general orientation of the government and the business sector of Shenzhen are market-driven, and so is Guangdong province. As such, the relatively good performance of Guangdong ETS pilot comes as no surprise. At the moment, we do not know exactly what is going on in Tianjin. Anyway, its low trading volume indicates that Tianjin is not ready as an active player in the carbon market in China. A plausible cause may have to do with firm ownership and industrial structure of the city for historical reasons. Deeper research is needed here.

4.2 Policy Recommendations

The ETS is purported to give China additional market-based tools to realize its Paris commitments. Since 2013, seven regional ETS pilots have been launched to test their validity in the context of China. According to China's overall roadmap of ETS construction, if these ETS pilots are successful during the first phase of construction (2011–2017), China would scale up them to the national level by the end of 2017 and begin the second phase (2017–2020) of construction of carbon market. In this context, our study represents an initial assessment of these regional ETS pilots, with an aim to provide lessons for the construction of a future unified ETS at the national level. We find that the success so far is limited and uneven amongst the seven regional ETS pilots, which warrants further investigation into the underlying causes.

The limited and uneven success of these regional ETS pilots calls for caution when thinking about nationwide expansion of carbon market. More systematic evaluation and deeper investigation should be conducted to pin down the underlying social and economic causes before decision of expansion to the national level is made.

To begin with, information transparency about actual carbon emissions is necessary. There, China urgently needs to build up a sophisticated system of monitoring, reporting, and verification (MRV) for corporate ESG (environment, social responsibility, and corporate governance), including the training of manpower. Indeed, a clean, reliable and robust MRV system is an integral part of the basic infrastructure of ETS. This is especially important when China launches a national ETS in the near term (in 2020), and in the medium- and long-term, integrates its ETS with international ETSs. In short, a unified digital MRV system should be treated as a prerequisite for current regional ETSs to scale up to a higher level in the future.[14] Here, regional and international cooperation and coordination with an aim to converge (rather than diverge) in the future is desired. This is especially the case were it also to reduce risks and facilitate long-term cross-border business investments. Besides, there are other two points to make at this moment, and they are as follow:

First, the limited achievement of ETS pilots triggers a more fundamental issue. That is, the role of local government officials. Indeed, it has been long argued that the incentive for local government officials has been the weakest link in the value chain of environmental protection in China. So far, the principal pursuit of Chinese local government officials is economic growth measured by GDP rather than long-term concerns of sustainable development. The ETS scheme will probably not be sufficient to alter the behavior of local government officials especially when carbon emission permits are distributed in a fashion of grandfathering. Indeed, there is initial evidence that those with auctions (e.g., Shenzhen, Guangdong, Hubei) performed better than those without (e.g., Chongqing, Tianjin) among the seven regional ETS pilots.

In the future, a priced carbon permit should be tested in ways similar to that of EU-ETS, where auction is allowed and the proportion of auction (relative to free allowances) gradually goes up over time. It might even be better if a carbon

[14] Understandably, the building of the data centers of digital MRV system for these regional ETS pilots would be an incremental process. But to think and plan ahead, these data centers as critical backbones of ETS networks should go green themselves. That is, they should not be driven by fossil energy but by renewable energy.

tax scheme be tested as an additional policy tool. France, for instance, imposes carbon tax of US$ 55/ton and so does Norway at US$ 64/ton.[15]

Second, the uneven success across the seven regional ETS pilots may reflect regional variations in firm ownership and industrial structure due to historical reasons. If it is true, caution is needed when one considers scaling-up regional ETS pilots to a unified national carbon market in China after 2017. It would seem that a big bang rather than a gradual approach is not appropriate. A better approach has to be sector-sensitive with phase-in mechanisms. More specifically, it would allow different phase-in periods for various sectors in the national ETS, while at the same time keeping some region-specific industries in the regional ETSs before the two systems—the national and the regional ETSs—converge incrementally and eventually join tracks in the long-run. The same approach should also apply to forging links with international ETSs in the long run.

Indeed, as we discussed earlier in the chapter, EU-ETS has phase-in designs, although the industrial structure of its member states is far more homogeneous than that of China. Otherwise, as unintended consequences, we may run the risk of making rich regions richer and poor regions poorer. In the meantime, fiscal subsidies or transfers across regions by the central government may serve as remedial measures to ameliorate the problem, one may argue. But ultimately, these measures contradict—or at least generate tensions with—the market principle of equality if and when China tries to join international ETSs in the long run. Beware of these tensions, as one moves forward.

That said and looking ahead, we would recommend that the overall approach to market-building of ETS should move with a vector from regional equilibrium to general equilibrium. In other words, given the experience gained thus far from these regional ETSs, what China should do years down the road is to build not just a simple and static model of "cap & trade" but a complex and dynamic model somewhat along the line from RICE to DICE model developed

---

[15] Of late the EU has started to consider imposing CBAM taxes (or cross border adjustment mechanism) to avoid carbon leakage. Although difficulties still exist at the moment in terms of data availability and exact calculations, if that happens, it would have immediate impact on Chinese exports, e.g., steel products. The long process of steelmaking (BF/BOF), which still dominates the Chinese steel sector (about 90%) is much more carbon intensive than the short process (EAF). EAF accounts for about 47% globally, excluding China. See Chapters 2, 3 and 4 of this book. *Ceteris paribus*, in case of imposing cross-border carbon tax, the global market share of Chinese steel exports would be squeezed.

by Nobel Laureate in economics Nordhaus (2000).[16] The complex model should simultaneously consider the economic, political, and institutional aspects of environmental intervention, along with international cooperation and coordination. Indeed, facing the enormous challenge of climate change, what is needed is not just breakthrough "hardware technologies" but also "sophisticated institutional technologies" to reduce GHG on this small village that we call Earth (Fu, 2009).

[16] RICE and DICE are a pair of models integrating climate change with economics developed by William Nordhaus, a 2018 Nobel Prize winner in Economic Sciences. RICE stands for regional dynamic integrated model of climate and the economy, whereas DICE stands for dynamic integrated model of climate and the economy.

Appendix

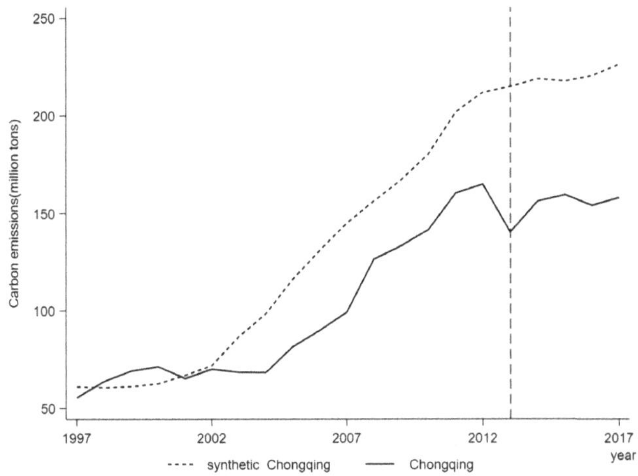

Fig. 11 CO_2 emissions between Chongqing and synthetic Chongqing

Acronym	Full spelling
CBAM	Cross border adjustment mechanism
CCER	China certified emission reduction
CCX	Chicago climate exchange
DICE	Dynamic integrated model of climate and the economy
ESG	Environment, social responsibility, and corporate governance
ETG	Emissions Trading Group
ETS	Emission-trading system
EU-ETS	European union greenhouse gas emission trading system
FYR	Five year plan
GHG	Greenhouse gas
KPI	Key performance indicators
MRV	Monitoring, reporting and verification
MSPE	Mean square percent error
NDRC	National development and reform commission
NSW	National Trust of Australia
RICE	Regional dynamic integrated model of climate and the economy
UNFCCC	United Nations framework convention on Climate Change

References

Abadie, A., & Gardeazabal, J. (2003). The economic costs of conflict: A case study of the basque country. *American Economic Review, 93*(1), 113–132.

Abadie, A., Diamond, A., & Hainmueller, J. (2010). Synthetic control methods for comparative case studies: Estimating the effect of california's tobacco control program. *Journal of the American Statistical Association, 105*(490), 493–505.

Abadie, A., Diamond, A. J., et al. (2015). Comparative politics and the synthetic control method. *American Journal of Political Science, 59*(2), 495–510.

Ando, M. (2015). Dreams of urbanization: Quantitative case studies on the local impacts of nuclear power facilities using the synthetic control method. *Journal of Urban Economics, 85*(C), 68–85.

Coase, R. H. (1960). "The Problem of Social Cost". *Journal of Law and Economics*, The University of Chicago Press, vol. 3, October, 1–44.

Duan, M., Tao, P., & Zhang, X. (2014). Review of carbon emissions trading pilots in China. *Energy & Environment, 25*, 527–550.

European Commission. (2003). Directive 2003/87/87/EC of the European Parliament and of the Council of October 2003 establishing a scheme for greenhouse gas emission allowance trading within the Community and amending Council Directive 96/61/EC.

European Commission. (2015). *EU ETS handbook*. European Commission.

Fu, J. (2009). *The Dao of the Wealth of Nations*. Peking University Press. p 373.

Kossoy, A., Oppermann, K., Platonovaoquab, A., et al. (2015). *State and trends of carbon pricing*. World Bank Group Climate Change, Washington DC: October 2015.

Mideksa, T. K., & Weitzman, M. L. (2019). Prices versus quantities across jurisdictions. *Journal of the Association of Environmental and Resouce Economisits., 6*(5), 883–891.

Liu, J., & Fan, Z. (2018). Research on the influence of industrial transfer from the property taxes: Empirical research from Chongqing and Shanghai(in Chinese). *China Industrial Economics, 11*, 98–116.

Nordhaus, W. D., & Boyer, J. (2000). *Warming the world: economic models of global warming*. The MIT Press (2000).

Pang, T., & Duan, M. (2016). Cap setting and allowance allocation in China's emissions trading pilot programs: Special issues and innovative solutions. *Climate Policy, 16*, 815–835.

Pigou, A. C. (1929). *The economics of welfare*. London: MacMillan and Co., Ltd.

Spence, A. M., & Weitzman, M. L. (1978). Regulatory strategies for pollution control. In: A. F. Friediaender (Ed.), *Approaches to controlling air pollution*. Cambridge MA: MIT Press.

Stavins, R. (2001). *experience with market-based environmental policy instruments*. Resource for the Future.

UNFCCC. (1998). Kyoto protocol to the United Nations framework convention on climate change. United Nations. Available at http//unfccc.int/Convkp/kpeng/kpeng.pdf.

Weitzman, M. L. (1974, October). Prices vs. Quantities. *The Review of Economics Studies, 41*(4), 477–491.

Weitzman, M. L. (2012). GHG targets as insurance against catastrophic climate damages. *Journal of Public Economic Theory., 14*(2), 221–244.

Weitzman, M. L. (2017). On a World Climate Assembly and the Social Cost of Carbon. *Economica, 84*(336), 559–586.

World Bank Group. (2018). *State and trends of carbon pricing 2018* (p. 2019). World Bank Group.

Zhang, J., Zhong, C., & Ming, Y. (2016). Did Olympic Games improve air quality in Beijing? Based on the synthetic control method. *Environmental Economics and Policy Studies, 18*, 21–39.

Design Options of ETS: How to Incentivize CCUS

6

Mengfei Jiang, Xi Liang, Ming Lei, Francisco Ascui, Qianguo Lin, Muxin Liu, and Li Wang

The urgency of rapid and ambitious climate action was underscored in IPCC Special Report on 1.5 °C. In the report, IPCC also emphasized that strong carbon price signals, coupled with further policy instruments, can support ambitious and cost-effective CO_2 emission reductions. As contracting states to the Paris Agreement look to implement their domestic climate policies, carbon markets are emerging and evolving to fill an important role in their strategic transition to a green economy. Worldwide, carbon markets, or emissions trading systems (ETS), have become a critical part of the policy response. China is no exception. As the previous chapter discussed, China rolled out 7 regional ETS pilots in 2013. Subsequently, as this chapter will discuss as a follow-up, China's national ETS was launched in 2017 and was expected to become operational around 2020.[1]

[1] At the time when this study went to the press, on July 16, 2021, China's national ETS officially went into operation, covering 2162 power plants and over 4 billion tons of carbon emissions annually. It thus became the largest of the 31 ETSs currently extant in the world. Whereas Beijing hosted China's national ETS center, Shanghai was tasked with trading, and Hubei with registration. The price of the first deal on that day was RMB52.78/tCO_2—much higher as compared to the average price of about RMB40/tCO_2 of China's 7 regional ETS pilots in past two years. It has been expected that China's national ETS will eventually cover petrochemical, chemical, building material, iron/steel, nonferrous, paper-making, and aviation industries as well.

M. Lei (✉)
School of Management, Peking University, Beijing, China

M. Jiang · X. Liang · F. Ascui · Q. Lin
University of Edinburgh Business School, Edinburgh, UK

M. Liu · L. Wang
UK-China (Guangdong) CCUS Center, Guangzhou, China

J. Fu et al. (eds.), *Climate Mitigation and Adaptation in China*,
https://doi.org/10.1007/978-981-16-4310-1_6

According to the International Energy Agency (IEA), the steel sector accounts for over 5% of global anthropogenic GHG emissions. China is the largest steel producer in the world, accounting for about 51% of global crude steel production in 2018. Currently, the steel sector accounts for over 10% of China's total domestic anthropogenic CO_2 emissions, making it the third largest source of CO_2 emissions in China—after power and cement. How best to decarbonize the steel sector in China in the years ahead? Essentially, there are basic four pathways to de-carbonize the iron/steel sector: (a) switching to zero-emission fuels (e.g., hydrogen), (b) improving energy efficiency, (c) structural adjustment (e.g., from long to short process), and (d) implementing carbon capture, utilization and storage (CCUS).[2] As pathway (a) would incur large costs to change the exiting outlay of the steeling making process, thus far China has largely relied on pathways (b) and (c); but the marginal effects of (b) and (c) have been on the decline. In this context, one would expect that pathway (d), i.e., CCUS, will increasingly become important in the years ahead.

Building upon previous discussions on carbon market in Chapter 5, the aim of this chapter is to further explore CCUS in connection with ETS. Specifically, the central question is: as China continues to build a national carbon market based on the experience of 7 regional ETS pilots, how could different design options of ETS incentivize CCUS application in the steel sector, given the fact that the current cost of CCUS has remained high in spite of cost reductions in recent years? To answer this question, we will put Chinese experience in a broader international perspective.

This chapter is organized into four sections. For international comparisons, Section 1 begins by briefly reviewing the key design elements of international experience. Section 2 then looks at the current designs and operations of ETS pilots in China Drawing heavily on EU experience, Section 3 introduces and analyses three different design options of ETS in connection with financing schemes of CCUS in the iron/steel sector. Conclusions follow in Section 4.

1 Section 1

1.1 Key Design Elements of ETS in International Perspective

Essentially, there are two generic approaches to emissions trading system (ETS) (see Fig. 1): cap-and-trade and baseline-and-credit systems. In cap-and-trade, a

[2] For more on CCUS, see Chapters 3 and 5 of this book.

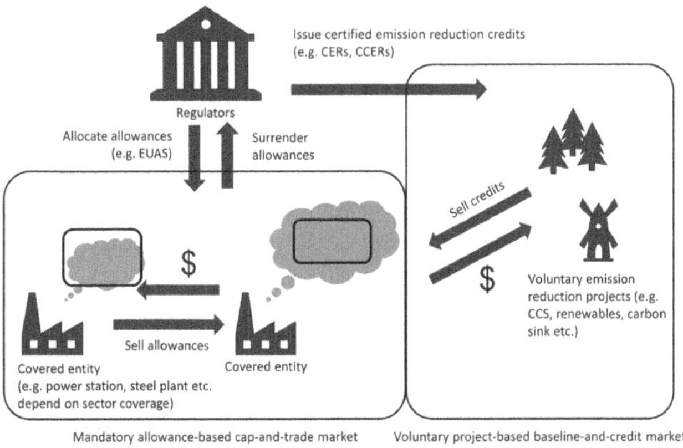

Fig. 1 Operation mechanism of ETS

regulator sets the total amount of emissions that can be emitted during a certain period (the cap) by issuing a limited number of allowances to emit. At the end of each compliance period, covered entities[3] Covered entities are defined as installations in EU-ETS and as enterprises in China's ETS must surrender allowances equal to their emissions during a certain period.[4] Covered entities could acquire allowances directly from the regulator (primary market) or from other participants by trading in the secondary market. Allowance allocation is the key to a successful carbon market (Xiong et al., 2015). Most ETSs around the world have started with free allocation but gradually transitioned to a higher percentage of allowances being auctioned.[5] Regulators usually apply one of two different allocation methods to allocate free allowance to covered entities: grandfathering and benchmarking.

[3] For more on CCUS, see Chapters 3 and 4 of this book.

[4] The total amount of emission allowances could be offset by certified emission reduction if there is a flexible mechanism for compliance.

[5] For example, in EU-ETS, the power sector started off with 95% free allocation of emission allowance in 2005 but gradually transitioned to 100% auctioning by 2013 (ICAP, 2019b). In China, the pilot ETS in Guangdong province required the power sector to purchase 3% of its emission allowances when the ETS started in 2013 (ICAP, 2019a).

Grandfathering is based on the enterprises' historical emissions or emission intensities, while benchmarking is based on a benchmark emission intensity at the sector or sub-sector level. A baseline-and-credit approach depends on there being some source of voluntary or regulated demand for offset credits from emission reduction projects. The Kyoto Protocol Clean Development Mechanism (CDM) was an international baseline-and-credit mechanism that allowed emission reduction projects in developing countries, including China, to earn offset credits known as Certified Emission Reductions (CERs) that developed countries could purchase and use to partially fulfil their Kyoto Protocol emission reduction targets.

2 Section 2

2.1 Current Designs and Operations of ETS Pilots in China

China started pilot emissions trading schemes in 2013, and the first seven pilot schemes (Beijing, Shanghai, Shenzhen, Tianjin, Chongqing, Hubei and Guangdong) have further developed their ETSs by expanding coverage, refining their allocation mechanisms, and introducing derivative products in the past few years. Many other provinces which were not included in the first seven pilot markets (such as Jiangsu, Xinjiang, Zhejiang provinces) have been interested in establishing their own schemes, and Fujian launched its own ETS in late 2016. By the end of April 2018, 222 $MtCO_2$ equivalents of allowances had been traded in the primary and secondary ETS markets in China, with a value of RMB 5 billion (circa. US$ 725 m) (World Bank Group, 2019).

Most ETSs in China only cover CO_2 emissions, apart from Chongqing, which covers six GHGs (see Table 1). All Chinese pilot ETSs cover both direct and indirect emissions. Covered emission sources in these pilots are defined as enterprises, rather than installations or facilities as in the EU-ETS (ICAP, 2019a). As discussed in the previous chapter, China is a country with significant regional diversities. Indeed, each province or city that participates in a pilot ETS has a different industrial structure and economic growth outlook, resulting in each pilot ETS having different sectoral coverage in order to achieve its target without adversely affecting economic growth projections (EYGM, 2014).

For example, Beijing, Shanghai and Shenzhen are commercial centers along the coast with relatively high GDP per capita, and carbon emissions from the building sector account for a significant proportion of emissions in these provinces or cities, therefore the sector is included in these pilots (Duan et al., 2014). Conversely, Hubei and Chongqing are located in central China with a relatively lower

Table 1 Sectoral coverage of pilot ETSs (2018)

Pilot ETS	Estimated coverage (%)	Sectors	Inclusion threshold	Number of firms	Allocation methods for steel sector
Beijing	45	Power and heat, petrochemicals, manufacturing (including iron and steel), building, public transportation and service	Total emissions: 5000 tCO_2/year	903 (5)	Grandfathering
Chongqing	50	Power and heat, electrolytic aluminum, ferroalloys, calcium carbide, cement, caustic soda, and iron and steel	Total emissions: 20,000 tCO_2/year	230 (n/a)	Grandfathering
Fujian	60	Power, petrochemical, chemical, building materials, iron and steel, nonferrous metals, paper, aviation, and ceramics	Energy consumption: 10,000 tce/year	255 (26)	Grandfathering
Guangdong	60	Power and heat, petrochemicals, manufacturing (including iron and steel), transportation (including aviation)	Total emissions: 20,000 tCO_2/year	249 (53)	Primary process: benchmarking; Secondary process and others: grandfathering

(continued)

Table 1 (continued)

Pilot ETS	Estimated coverage (%)	Sectors	Inclusion threshold	Number of firms	Allocation methods for steel sector
Hubei	45	Power and heat, petrochemicals, chemicals, manufacturing (including iron and steel, nonferrous metals, cement etc.), food and beverage	Energy consumption: 10,000 tce/year (in any year between 2014 and 2016) 60,000 tce/year (in any year between 2010 and 2011)	344 (23)	Grandfathering
Shanghai	57	Power and heat, petrochemicals, chemical, iron and steel, nonferrous metals, building, transportation (including aviation, shipping etc.) and commerce	Total emissions: 10,000 tCO_2/year 100,000	289 (4)	Grandfathering
Shenzhen	40	Power and heat, water-supply, manufacturing (including iron and steel), building and public transportation	Total emissions: 3000 tCO_2/year	794 (2)	Grandfathering
Tianjin	55	Power and heat, petrochemicals, chemicals, iron and steel, oil and gas production	Total emissions: 20,000 t CO_2/year	107 (29)	Grandfathering

Source ICAP (2019b)

GDP per capita but higher GDP growth rates. Accordingly, these pilot ETSs are more focused on heavy industrial sectors. Major emitting sectors such as power and heat, petrochemicals and manufacturing are covered in all pilots.

Table 1 summarizes the sectoral coverage of the eight pilot ETSs in China.

The iron and steel sector is covered by all pilot ETSs. Allowances are mainly allocated for free to enterprises in the sector, although auctioning may take place where it is considered appropriate by the ETS authorities of Shenzhen, Beijing, Guangdong and Fujian for the purpose of adjusting the market price (Pang & Duan, 2016). All pilots have adopted the historical emissions "grandfathering" method to allocate allowances to covered iron and steel enterprises, except for the Guangdong pilot ETS, which has used a benchmarking method for iron and steel enterprises equipped with primary process steel making. (Primary process (or primary route) refers to basic oxygen steel making process that uses).

Building on the successful experience of regional ETS pilots since 2013, China's National Development and Reform Commission (NDRC) announced the official launch of a national ETS in December 2017, aiming to contribute to the effective control and gradual reduction of carbon emissions in China and the achievement of low carbon and green development. The national ETS regulates about 1,700 enterprises from the power sector, which emit over 26,000 tons of CO_2 per year or consume over 10,000 tons of standard coal equivalent (tce) per year (ICAP, 2019b). More than 3 $GtCO_2e$ are covered in the initial phase, which accounts for approximately 30% of China's national emissions.

China's national carbon market is planned to start operation in 2020, although there is a delay due to the climate change department being moved from NDRC to the Ministry of Ecology and Environment (MEE) in 2018 (Zaman & Yang, 2019). The provisions for the launch and gradual development of the national ETS are laid out in the Work Plan for Construction of the National Emissions Trading System (Power Sector). The Work Plan was approved by the State Council in late 2017 (NDRC, 2017a). A phased transition in the Chinese ETS pilots is foreseen by the Work Plan. In the short term, the existing regional ETS pilots are expected to operate in parallel with the national carbon market, covering existing power and non-power sectors. Over the medium to long term, however, they are expected to be integrated into the national carbon market, once it is fully operational.

The power sector generally has higher quality of emissions data and it is therefore the first sector to be included in the national ETS. On 15th December 2017, the NDRC issued a notice (NDRC, 2017b) requiring reporting and verification of 2016 and 2017 carbon emissions data, as well as the development plans for future monitoring, reporting and verification (MRV) work in eight industrial sectors. These eight industrial sectors are: petrochemicals, chemicals, building materials, iron/steel, nonferrous metals, papermaking, electricity, and aviation. The coverage of these industrial sectors in the notice signals that the long-run plan is to integrate these sectors into the national ETS or another carbon pricing or regulation scheme.

Domestic project-based offset credits are allowed in China's pilot ETSs, meaning that emissions can be reduced by sources that are not covered by the schemes and that are not located in the jurisdictions. The use of offsets is limited to 1% of annual allowances in the Shanghai ETS pilot, 5% (Beijing and Fujian ETS pilots), 8% (Chongqing ETS pilot) and 10% (Guangdong, Hubei, Shenzhen and Tianjin ETS pilots) (ICAP, 2019b).

Table 2 below provides an overview on how the use of offset credits is regulated in the eight regional ETS pilots. None of these regional ETS pilots allows

Table 2 Offsets regulations in China's pilot ETSs

Pilot ETS	Eligible type of offset credits	Offsets limit (% of annual allowances)	Other requirements
Shenzhen	CCERs	10	Credits from hydro projects not allowed
Shanghai	CCERs	1	Credits from hydro projects not allowed
Beijing	CCERs	5	n/a
Guangdong	CCERs and Pu Hui Certified emission Reduction (PHCER)	1.5 million offsets	At least 70% from projects in Guangdong province Credits from hydro projects not allowed
Tianjin	CCERs	10	Credits from hydro projects not allowed
Hubei	CCERs	10	Credits have to be generated from rural biogas and forestry projects located in poor areas of Hubei
Chongqing	CCERs	8	Credits from hydro projects not allowed
Fujian	CCERs and Fujian Forestry Certified Emission Reduction (FFCER)	10	n/a

the use of international offsets from outside China. To avoid potential carbon leakage[6] Hubei ETS, with a relatively low abatement cost, requires that all CCERs must be generated from projects within Hubei province, while Guangdong ETS requires at least 70% of own-produced CCERs for the same reason (IETA, 2014, ICAP, 2019b). With concerns about the impact of CCERs on the ETS pilot markets and an upgrade of the current CCER system, the CCER project registration process was suspended since March 2017 (NDRC, 2017a).

At the 13th National People's Congress (NPC) in March 2018, the mission of combating climate change and reducing emissions was transferred from the NDRC to the MEE, which is likely to promote synergies between the development of a national carbon market and efforts to reduce air pollution (Slater et al., 2018). The Climate Change Department within the Ministry of Ecology and Environment will operate the national carbon market jointly with provincial governments. In April 2019, MEE started a consultation on the Interim Regulations for the Administration of Carbon Emission Permit Trading (MEE, 2019). The Interim Regulations set the rules for allowance allocation and trading in the regions, and mandate local governments to collect emissions data and regulate the emission trading market.

3 Section 3

3.1 Options for China's ETS to Incentivize Steel Sector CCS

The purpose of an ETS is to put a price on emissions and so incentivize emission reductions in a cost-effective way. It is therefore relevant, albeit tentative, to consider how China's national ETS could be used to incentivize CCUS in the steel sector. In this section we discuss three tentative approaches to this: (1) allowing CO_2 stored through CCUS to be regarded as "not or low emitted" for the purposes of ETS compliance; (2) allowing offsets from iron/steel CCUS projects to be used for compliance in China's ETS; and (3) incentivizing CCUS with finance from auctioning of ETS allowances or directly awarding allowances for steel plants with CCUS.

[6] "Carbon leakage refers to domestic carbon emissions that are displaced by relatively cheaper imported products, mainly from countries or regions without carbon constraints, rather than reduced as anticipated because of the mitigation of unilateral regional climate change" - WANG, X., TENG, F., WANG, G., ZHOU, S. & CAI, B. 2018. Carbon Leakage Scrutiny in ETS and non-ETS Industrial Sectors in China. *Resources, Conservation and Recycling*, 129, 424–431.

3.2 Regarding CO_2 Stored via CCS as "Not or Low Emitted"

One option to incentivize CCUS implementation in the steel sector in China would be to include emissions from the sector within the scope of the national ETS, while simultaneously allowing CO_2 stored through CCUS to be regarded as 'not or low emitted' for the purposes of ETS compliance. Within the option, regulators would need to consider the amendment of existing legislation, including potential changes in its coverage, allowance allocation method and MRV system, and introduce new rules concerning the capture and geological storage of carbon dioxide.

ETS Coverage

This option depends on the iron/steel sector being covered by the ETS. As previously discussed, the pilot ETSs in China all include the iron/steel sector, but the national ETS does not. Therefore, new legislation would be required to change the scope of the national ETS. This would need to specify which sources of emissions in iron and steel plants are covered within each entity's defined scope. In the EU-ETS, sources are defined in terms of activities and thresholds (where the intention is not to include 100% of emissions from an activity).

For example, all coke making, sintering and palletization plants are included, regardless of capacity; whereas production of pig iron or steel is included where production output is over 2.5t/day, and processing is included where combustion units over 20MWth input capacity are operated (European Commission, 2003). This leads to approximately 75% of the sector's emissions being covered in the EU-ETS (Dechezleprêtre et al., 2018). In terms of covered gases, it is likely that only carbon dioxide emissions from the iron/steel sector would be covered, in accordance with the coverage for the power sector in the national ETS. This would not leave a significant amount of GHG emissions unregulated, as relatively small quantities of other GHGs are emitted from iron/steel production (principally small amounts of methane and nitrous oxide from the combustion of coal) (Beer et al., 2003).

Iron and Steel GHG Accounting

GHG emissions from an enterprise are conventionally divided into three scopes. Scope 1 refers to direct emissions from owned or controlled sources while Scope 2 refers to indirect emissions from the generation of purchased energy. Scope 3 includes all other indirect emissions (not included in scope 2) that occur in the supply chain of the reporting enterprise, both upstream and downstream. In the EU-ETS, covered entities are liable only for their Scope 1 emissions. In China, however, the existing *Guidelines for Accounting and Reporting Greenhouse Gas Emissions: Iron and Steel ProductionEnterprises in China* (NDRC, 2013a) require information on both direct emissions (Scope 1) from combustion of fossil fuel and consumption of solvent flux, carbonaceous materials and electrodes, and indirect emissions (Scope 2) from net purchased power and heat. Further details on the current accounting method for GHG emissions in the iron and steel sector in China can be found in Appendix 1.

Allowance Allocation

As previously observed, China's ETS pilots have allocated free allowances to the iron and steel sector based on a grandfathering (historical data) approach, rather than benchmarking (based on activities) (see Table 1). A generic calculation of the allowances allocated to an iron and steel enterprise in China's ETS pilots can be given as Formula 1 below:

$$A = B \times F_m \times Q \tag{1}$$

where.

A Total allocated allowances to covered iron and steel entity.

B Historical emissions or emissions intensity.

F_m Emission reduction factor.

Q Crude steel production.

Although benchmarking is more conducive to encouraging advanced production processes, it is relatively difficult to determine suitable benchmarks for the iron and steel sector, due to its highly diverse product portfolio and complex production processes, as well as the lack of adequate emissions data. Nevertheless, after lessons learned from the initial two phases, the EU-ETS has switched its allocation method for industry and heating sectors in the third phase (2013–2020) from grandfathering to benchmarking (European Commission, 2015a).

CCS Recognition

Meanwhile, CCUS needs to be recognized legally so that emissions captured, transported and stored would not be treated as emitted. This occurred in a phased manner in the EU-ETS. From the start of Phase II in 2008, Member States had the option of including CCUS in the EU-ETS, subject to the Commission's approval. Under the revised EU-ETS Directive 2009/29/EC, CCUS was given fuller recognition within the scheme, which took effect from Phase III in 2013

(European Commission, 2009a). The most outstanding element of the EU regulation on CCUS was a separate Directive for the geological storage of CO_2, which regulates the risks of CO_2 storage (European Commission, 2009b) and allows each Member State to formulate its regulatory approach to comply with the EU requirement (European Commission, 2011). In addition, each Member State could formulate its own policies for incentivizing CCUS demonstration projects, subject to State Aid approval by the European Commission.

In terms of the impact on CCUS financing, the option of allowing CO_2 stored via CCUS to be regarded as 'not or low emitted' for the purposes of ETS compliance would boost confidence for industries to demonstrate CCUS, but the actual financial support provided would be very limited. Given that average price expectations in the national ETS are RMB 54/ton (US$7.7/ton) in 2020 and RMB 98/ton (US$ 14/ton) in 2025 (Slater et al., 2018), and the cost of CO_2 avoidance for a 0.5 million ton/year capacity CO_2 capture project with pipeline transport and storage in an offshore saline formation is estimated at RMB 443/tCO_2 (US$ 63/tCO_2) (Liang et al., 2019), this option would only contribute a small percentage of the required incentives to demonstrate CCUS. This finding is similar to the recommendation of a study by ENGO Network (2013) which suggested that an ETS could only form part of a financing package for CCUS. However, the cost estimate was based on offshore saline formation storage: if CCUS in China was deployed with onshore enhanced oil recovery (EOR) where the abatement cost could be less than US$20 per ton CO_2 avoided (NCEPU, 2019), the carbon price in the ETS could provide a more substantial part of the incentives required to enable CCUS, although still not the entire amount.

3.3 Allowing Offsets from Steel CCUS Projects in China's ETS

Allowing CCUS to be included in project-based baseline-and-credit emissions trading is a second option for incentivizing iron/steel CCUS development in China. This would involve allowing CCUS projects to generate offsets which could be sold to participants in the ETS, who would be allowed to use them to meet their compliance obligations.

So far none of the eight regional ETS pilots have allowed emission reduction offset credits from CCUS to be eligible for compliance. On 28th December 2018, the NDRC, together with eight other government institutions including the MEE, issued the "Action plan on the establishment of a market-oriented and diversified ecological protection compensation mechanism" (NDRC, 2018). The Action Plan mentions the review and improvement of the CCER offset scheme and specifically prioritizes the use of forestry carbon sink projects as offsets for the national ETS, but CCUS is not explicitly discussed in the document.

In terms of allowing CCUS projects in the iron/steel sector to generate offset credits, two options are available: either domestic legislation to allow such projects to generate credits that would be unique to China's national ETS, or making use of the international legal framework for recognizing emission reductions in developing countries under the Kyoto Protocol Clean Development Mechanism (CDM). As previously discussed, the CDM allows emission reduction projects in developing countries to generate internationally-recognized credits known as Certified Emission Reductions (CERs) (Dixon et al., 2013). In December 2011, at the meeting of the UNFCCC in Durban, South Africa, CCS was formally included in the list of project activities that may be eligible as CDM project activities (Brander & Ascui, 2018). However, at the present time, no carbon accounting methodologies for CCS have been put forward for approval under the 2011 agreement. Until such a methodology, applicable to CCUS in the iron/steel sector in China, is put forward and approved under the CDM modalities and procedures, it will not be possible to generate internationally recognized CERs from such a project. Therefore, the option of domestic legislation allowing projects to generate a China-specific offset credit is currently considered more feasible.

Such legislation would need to establish an administrative procedure for approving a GHG accounting methodology for CCUS projects in China. For project-based GHG accounting, emission reductions are quantified by comparing measured project emissions against a forward-looking, counter-factual baseline scenario or reference level (World Resources Institute and World Business Council for Sustainable Development, 2005). In addition, any increase in emissions elsewhere that is measurable and attributable to the project, yet not within the project boundary, is denoted 'leakage' and also deducted from the calculated emission reductions (Brander & Ascui, 2018). Therefore, a general formula for project-based baseline-and-credit accounting is shown in Formula 2 below:

$$ERy = BEy - PEy - LEy \tag{2}$$

where:

ERy Emission reductions in year y (tCO_2).

BEy Baseline emissions in year y (tCO_2), representing emissions from iron/steel enterprise that would have entered atmosphere if not for CCUS project

PEy Project emissions in year y (tCO_2), representing actual GHG emissions from iron/steel enterprise after implementation of CCUS project, including carbon capture, transport pipelines, and storage sites.

LEy Leakage emissions in year y (tCO_2).

An important challenge for project-based GHG accounting is identifying and characterizing the baseline scenario and estimating the baseline emissions (*BEy* in Formula 2). A baseline is a hypothetical situation that represents the conditions most likely to occur in the absence of the CCUS project. The baseline should cover all emissions within the project boundary. For an iron/steel CCUS project, establishing a baseline for each component is suggested (Brander & Ascui, 2018). Existing actual or historical emissions would be a suitable baseline approach for a CCUS project that involves the addition of CCUS to an existing iron/steel plant. The baseline for retrofit CCUS project would involve the continued operation of the existing CO_2 source facility, but without CCUS, such that produced CO_2 is vented to the atmosphere. For a new build CCUS project, the baseline could correspond with the common practice in the area and the most economic option available, or an emissions performance standard that best represents project circumstances (McCormck, 2012).

After a project is implemented, emissions need to be monitored and reported. Emissions from a full chain CCUS project would comprise all sources of emissions within the project boundary, which would include at a minimum: The installation where CO_2 captured;

- Any treatment facilities;
- Transportation equipment;
- Injection facilities; and.

Subsurface components, including the storage site and potential sources of seepage, as well as the vertical and lateral limits of the geological storage site that are expected when the CO_2 plume stabilizes over the long term.

Calculating project emissions involves measuring CO_2 emitted during capture, compression, transportation, injection and storage (and recycling of CO_2 if applicable) and deducting CO_2 permanently stored by the project. Emissions generated from the additional energy required for capture (known as the 'energy penalty') should be taken into consideration. Any seepage emissions (CO_2 that escapes from a geological reservoir into the ocean or atmosphere) should also be include in project emissions.

In terms of the impact on investment, the price of CCERs is generally lower than the price of allowances, so the financial impact of allowing iron and steel

CCUS projects to generate CCERs is less than under approach 1. The cost of monitoring, reporting and verification (MRV) of emission reductions at the project level is also generally higher than the equivalent costs under approach 1. There have not yet been any CCUS projects, anywhere in the world, financed solely through the CDM or similar offset mechanisms. Nevertheless, allowing CCUS projects to generate offset credits that can be used within China's ETS is an option that could be considered if approach 1 is infeasible (for example, if it is decided that the iron/steel sector will not be covered by the national ETS, even in the next phase). While the revenue from offset credits would not be sufficient on its own to incentivize CCUS projects, it could boost confidence for investors in CCUS and help establish the MRV system for the iron and steel sector.

3.4 Auctioning ETS Allowances for Iron/Steel Plants with CCUS

The third approach involves using the revenue generated by the auctioning of allowances to support CCUS development and demonstration in the iron/steel sector. The government could establish a funding pool to provide dedicated support for financing CCUS demonstration projects. There has been some experience of this already in China's pilot ETSs. The Guangdong ETS has attempted to use approximately RMB 1 billion (US$140 million) in auctioning revenue to finance innovative low-carbon technologies, but the provincial government finance department has been reluctant to approve the expenditure. An alternative approach could be to directly award allowances (taken from the amount to be auctioned) to companies building CCUS projects. The latter method could be easier to implement as it would not require finance department approval, but it would expose the project developer to price risk as the value of their allowances could change over time. Under a direct funding approach, the government rather than the project developer would be exposed to price risk.

The closest international reference is the EU-ETS New Entrant Reserve (NER). In 2009, the European Commission (EC), European Investment Bank (EIB) and EU Member States agreed to set aside 300 million allowances for subsidizing CCUS and installations of innovative renewable energy technologies, hence the fund's name 'NER 300' (European Commission, 2017). With carbon prices heading toward EUR 30/ton in 2008, it was hoped that up to EUR 9 billion would be raised. However, prices subsequently crashed, and it was not until late December 2012 that the European Commissioner for Climate Action Connie Hedegaard finally announced the outcome of the first NER300 funding round—but could only award EUR 1.2 billion, due to the drop in allowance price, to 23 innovative renewable energy projects across Europe. Not a single CCUS project was funded, and due to the long selection process, the CCUS projects that had applied for NER300 funding missed their most ideal timing to receive investment under legislation for Phase 4 (2020–2030) of the EU-ETS, which became effective in April 2018, at least 400 million allowances (called NER400) will be reserved from 2021 onwards to create an Innovation Fund for demonstration projects of renewable energy and environmentally-safe CCS/CCUS. In addition, a further 50 million unallocated allowances from the Market Stability Reserve, together with remaining unallocated funds from the existing NER 300 Program, will be made available to support eligible projects.

To enable this approach, three key elements need to be considered for regulating ETS funds. First, regulators need to define the financing conditions, including decisions on how much support for projects is available, different ways the funds can be disbursed, i.e. as grants or loans for instance, whether a co-financing rate exists, and any other conditions that should be applied. Second, clear eligibility criteria and a transparent selection process need to be defined. Typically, a selection process involves some kind of ranking of eligible proposals so that they can be selected for funding in rank order, until the available funding limit is reached or there are no further eligible proposals. Third, the responsibilities of different institutions with regards to the implementation of the funding program need to be defined. Such responsibilities include, but are not limited to, issuing calls for proposals, monetizing ETS emission allowances, and assessing the technical and financial viability of projects.

Auctioning revenue of ETS allowances could be a viable option to finance the first few CCUS demonstration projects in the iron/steel sector in China, assuming that the scale of ETSs in China could generate sufficient revenue to support financing CCUS projects. Auctioning revenue of ETS allowances is also the most flexible option, and this option can also be combined with the other options. For example, the steel sector could be brought into the ETS as per option 1, with revenue from auctioning allowances to the power sector being recycled to offset the impact on the steel sector's international competitiveness. It could also be combined with project-based crediting, by providing successful projects with additional funding which could be in the form of a fixed amount, or a variable amount, e.g., based on meeting the difference between the current carbon price and the abatement cost, or a multiple of the carbon price.

Two potential limitations of this approach are the uncertainty in the carbon price, and the need for agreement by relevant government finance departments to establish and allocate special funds, as discussed earlier. Policymakers in China could consider directly awarding allowances to steel companies building CCUS projects to avoid the financial approval process, but this method should be temporary. Finally, a limitation of this approach is that it could not be scaled to support every possible steel sector CCUS project, but could potentially incentivize early demonstration projects with adequate financial support, which would enable costs of next-generation projects to fall by learning.

The amount and type of MRV required for CCUS projects funded in this way would depend on the nature of the selection process and the objectives of the funding program. For example, if the objective is purely to demonstrate a carbon capture technology, then it may not be necessary to set in place a system for monitoring long-term storage. However, in most circumstances it would still be desirable to have robust measurement and reporting of the actual emission reductions (or potential emission reductions, in the case of capture-only projects) resulting from the funded projects, using a consequential accounting methodology as outlined under option 2, including MRV for geological storage and reversal of storage, liability, insurance and EOR.

In addition, the conditions of the funding allocation may set certain conditions which have to be met, such as meeting a certain timeline, capture rate or total amount of CO_2 stored, and these parameters would need to be monitored.

4 Section 4

4.1 Think Ahead

Carbon emission trading system (ETS) is a market-based mechanism that can help to achieve emission reduction targets in a cost-effective fashion. In this chapter, we have explored different design options of ETS in connection with CCUS. Specifically, to better incentivize CCUS development in China, we have *tentatively* [emphasis added] discussed three options, and with a focus on the iron/steel sector. These design options are "tentative", in view of the fact that while China's regional ETS pilots have covered the iron/steel sector, China's national carbon market, launched in late 2017, has yet to cover the iron/steel sector. But we believe that the exploration is not futile, because the logic could apply to other sectors (e.g., the power sector).

For the time being, the national carbon market only covers the power sector and the existing regional ETS pilots are expected to operate in parallel with the national carbon market. Over the medium- to long-term, however—along with petrochemicals, chemicals, building materials, nonferrous metals, papermaking, electricity and aviation, the iron/steel sector is expected to be integrated into the national carbon market (NDRC, 2017b, 2017c). So it is relevant to think and plan ahead.

4.2 Concluding Remarks

To recap, there are three tentative design options, and they are as follows: The first option is to allow CO_2 stored through CCUS to be regarded as "not or low emitted" for the purposes of ETS compliance, such that CCUS-fitted or retrofitted iron/steel plant can realize emission reductions at the time of performance. This option would allow CCUS-fitted plants to generate revenues by selling spare allowances, or not having to purchase additional allowances in the carbon market.

The second option is to adopt a project-based baseline-and-credit approach. In this option, the national ETS covered emitters could purchase offset credits from CCUS-fitted or retrofitted iron/steel plants and use these purchased credits to meet their ETS compliance obligations. The third option is to use revenues generated by auctioning off emission allowances to support development and deployment of CCUS technologies in the iron/steel sector.

In comparative perspective, the first option has more straightforward MRV requirements than the second, but requires a change to enlarge the scope of the national carbon market to include the iron/steel sector, whereas the second option could be implemented relatively easily. Yet, for the first two options work effectively, the overall ETS cap has to set at a level ambitious enough to achieve a sufficiently high price, which means, at least theoretically, all cheaper abatement options would have to be exhausted before CCUS in the iron/steel sector become the marginal price-setting option. In other words, CCUS would need to compete with all other abatement options featuring lower costs. In reality, however, the current low prices in China's ETSs, regional or national, are unlikely to be able to support a CCUS project in the iron/steel sector, in the absence of additional subsidies from other sources.

Thus, by comparison, the third option offers a more promising way to support early-stage CCUS projects in the iron/steel sector. But even for this option to work, it must be endorsed by relevant government departments and detailed rules and procedures need to be developed and established. That said, standing alone, the third option is the most flexible options, but it is not mutually exclusive with the first or the second option, and indeed can be combined with either of them. It is therefore recommended if a staged approach is taken. In strategic sequencing, the third option, if rolled out first, could potentially lever much stronger support than the other two options in the short-term. Overtime, when the carbon price becomes sufficiently high and the long-term price signal becomes strong and stable,[7] other two options would become sufficient on their own. In any case, a robust MRV system as well as a sound regulatory framework would be a pre-condition for all three options to work effectively.

Table 3 summarizes the opportunities and challenges that each option may bring to CCUS in the iron/steel sector in China.

Finally, it is worth noting that while the sectoral cut of this study is the iron/steel sector (see Appendix 1), the logic of the three different design options of ETS to better incentivize CCUS applications could also apply to the power sector. The power sector is the first sector covered by China's national carbon market, which accounts for the lion's share (about 45–50%) of China's CO_2 emissions. Note also that, as discussed in Chapters 1 and 2, currently most of China's CCUS pilots are in the power sector. As a novel and emerging technology on industrial scale to directly reduce carbon emissions, CCUS is currently included in China's mid- to long- term development plan (MOST, 2013; State Council, 2006). Although the NDRC has provided policy support for CCUS demonstrations (NDRC, 2013b), there is not yet any dedicated regulatory framework in place to ensure the legal and commercial certainty for CCUS projects in China.

That notwithstanding and in view of China's mid- to long-term development plan of green growth, this study has looked ahead in exploring different design options of China's ETSs in connection with the financing of early-stage CCUS demonstration projects. And so doing, it has brought us back to the nexus of this study, i.e., the synergy between policy, technology, and market to push forward the frontier of inclusive and sustainable growth.

---

[7] Price stability is important. Otherwise, the funding required for CCUS is uncertain: more would be required if the allowance price falls, while at the same time, auction revenues would also fall. To counteract this effect, it is proposed that the volume of allowances set aside for the fund could be flexible, thus operating in a similar way to the Market Stability Reserve in the EU-ETS, withdrawing more allowances from the market when the price is low, and less when the price is high. This would have the beneficial effect of creating greater price stability in the ETS as a whole.

Table 3 Comparison of 3 design options of ETS

Options	Opportunities	Challenges
(1) Including the iron/steel sector in the national ETS and allowing CO_2 stored through CCUS to be regarded as 'not or low emitted' for the purposes of ETS compliance obligations	– Steel plants with CCUS could realize emission reductions at the time of performance, generating revenue by selling spare allowances or avoiding extra costs or not having to purchase from the carbon market – MRV requirements are relatively straightforward compared with option 2, as all steel plants would already have to monitor and report their emissions, so additional MRV would only apply to CCUS components	– Carbon prices are unlikely to be high enough to incentivize CCUS implementation on their own, as all abatement options in both the power sector and the steel sector that are cheaper than CCUS would be exploited first – A free allocation approach based on historical emissions would tend to have less effect to motivate iron/steel enterprises to adopt advanced carbon abatement technology such as CCUS
(2) Allowing offsets from iron/steel CCUS projects in China's ETSs	– The option enables sectors not covered by an ETS to generate revenues from reducing carbon emissions – This option does not require a major change to the scope of the national ETS	– Offsets will only be used for compliance within an ETS when their price is lower than the prevailing price of allowances in the ETS. Therefore, the likely revenue is lower than option 1 – The transaction costs of option 2 are also potentially higher than option 1
(3) Incentivizing CCUS with finance from auctioning of ETS allowances or directly awarding allowances to steel plants fitted with CCUS	– The most flexible option, which can also be combined with the other options	– Market price fluctuations would create uncertainty over the actual budget available for disbursement – Setting up new funds requires government finance department approval – CCUS might need to compete with other demands for such funds

Appendix 1

Table A1: GHG accounting guidelines for iron/steel industry in China (NDRC, 2013a)

	Calculation formula	Notes
Eco_2 (Total CO_2 emissions in steel-making process)	$Eco_2 = E_{combustion} + E_{power\&heat} - E_{sequestration}$	Eco_2 presents total CO_2 emissions; *Ecombustion* refers to CO_2 emissions of all net consumed fossil fuel combustion activities; *Eprocess* represents CO_2 emissions from industrial production processes; *Epower&heat* is CO_2 emissions corresponding to net purchased power and heat by enterprise; *Esequestration* is implicit CO_2 of carbon-sequestered products (e.g., crude steel)
$E_{combustion}$ (CO_2 emissions of all net consumed fossil fuel combustion activities)	$E_{combustion} = \frac{44}{12}\sum_{i=1}^{n}(AD_i \times CC_i \times OF)$	*ADi* refers to activity level of fossil fuel consumed within accounting and reporting period (GJ); *CCi* refers to carbon content per unit of calorific value of fossil fuel (iC/GJ); *OFi* is the rate of carbon oxidation of fossil fuel (%)

(continued)

Table A1: (continued)

	Calculation formula	Notes
$E_{process}$ (CO_2 emissions from industrial production processes)	$E_{process} = E_{solvent} + E_{electorde} + E_{rawmaterials}$ $= \sum_{i=1}^{n} (P_i \times DX_i \times EF_i) + P_{electrode} \times EF_{electrode}$ $+ \sum_{i=1}^{n} M_i \times EF_i$	*Esolvent* is CO_2 emissions from solvent consumption; *Eelectrode* is CO_2 emissions from consumption; *Eraw materials* is CO_2 emissions from consumption of other carbonaceous materials such as pig iron, ferroalloy and direct reduced iron; *Pi* is net consumption of solvent; *DXi* is average purity of solvent (%) *EFi is* CO_2 emission factor of solvent (tCO_2/t); *Pelectrode* is amount of electrode consumed for EAF-based steel making (t); *EFelectrode* is CO_2 emission factor for electrode consumed; *Mi* is purchased amount of carbonaceous materials; *EFi* is CO_2 emission factor for purchased carbonaceous materials(tCO_2/t)

(continued)

Table A1: (continued)

	Calculation formula	Notes
$E_{power\&heat}$ (CO_2 emissions corresponding to net purchased power and heat by enterprise)	$E_{power\&heat} = AD_{power} \times EF_{power} + AD_{heat} \times EF_{heat}$	*ADpower* and *ADheat* are net purchased electricity (MWh) and heat (GJ) amount within accounting and reporting period; *EFpower* and *EFheat* are CO_2 emission factors for electricity (tCO_2/MWh) and heat (tCO_2/GJ)
$E_{sequestration}$ (Implicit CO_2 emissions of carbon-sequestering products)	$E_{sequestration} = \sum_{i=1}^{n} AD_{sequestration} \times EF_{sequestiration}$	$AD_{sequestration}$ is output of carbon sequestering product (t); $EF_{sequestration}$ is emission factor for carbon-sequestering product (tCO_2/t)

Acronym	Full spelling
AAUs	Assigned Amount Units
CCER	Chinese CER
CCS	Carbon capture and storage
CCUS	Carbon capture, utilization and storage
CDM	Clean Development Mechanism
CERs	Certified Emission Reductions
CO_2	Carbon dioxide
CO_2e	Carbon dioxide equivalent
EC	European Commission
FFCER	Fujian Forestry Certified Emission Reduction
EIB	European Investment Bank
EOR	Enhanced oil recovery
ETS	Emissions trading scheme
ERUs	Emission Reduction Units
EU ETS	EU Emissions Trading System
GDP	Gross Domestic Product
GHG	Greenhouse gas
IET	International Emission Trading
IPCC	Intergovernmental Panel on Climate Change
JI	Joint Implementation
MEE	Ministry of Ecology and Environment
Mtce	Million tons of coal equivalent
NAP	National Allocation Plans
NDC	Nationally Determined Contribution
NDRC	National Development and Reform Commission
PHCER	Pu Hui Certified Emission Reduction
SO_2	Sulphur dioxide
Tce	Ton of coal equivalent
UNEP	United Nations Environment Program

References

Beer, J. D., Harnisch, J., & Kerssemeeckers, M. (2003). *Greenhouse gas emissions from iron and steel production*. ECOFYS.

Brander, M. & Ascui, F. (2018). Consequential GHG assessment of carbon capture, utilization and storage for a steel plant in China. *BHP Carbon Capture, Utilization and Storage in China's Iron and Steel Sector Project.* University of Edinburgh. COASE, R. H. 1960. *The problem of social cost*.

Dechezleprêtre, A., Nachtigall, D. & Venmans, F. (2018). *The joint impact of the European Union Emissions trading scheme on carbon emissions and economic performance*. OECD Working Papers.

Dixon, T., Leamon, G., Zakkour, P., & Warren, L. (2013). CCS projects as Kyoto protocol CDM activities. *Energy Procedia, 37*, 7596–7604.

Duan, M., Tao, P., & Zhang, X. (2014). Review of carbon emissions trading pilots in China. *Energy & Environment, 25*, 527–550.

EEA. (2011). Air pollution impacts from carbon capture and storage (CCS). Copenhagen. EU

MERCI. (2018). *Technical analysis-iron and steel sector* (NACE C24.1–24.2–24–5).

European Commission. (2003). *Directive 2003/87/EC of the European Parliament and of the Council of 13 October 2003 establishing a scheme for greenhouse gas emission allowance trading within the Community and amending Council Directive 96/61/EC* (Text with EEA relevance).

European Commission. (2009a). Directive 2009/29/EC of the European Parliament and of the Council of 23 April 2009. *Official Journal of the European Union*. Strasbourg.

European Commission. (2009b). *Directive 2009/31/EC of the European Parliament and of the Council of 23 April 2009 on the geological storage of carbon dioxide and amending Council Directive 85/337/EEC, European Parliament and Council Directives 2000/60/EC, 2001/80/EC, 2004/35/EC, 2006/12/EC, 2008/1/EC and Regulation (EC) No 1013/2006* (Text with EEA relevance). In UNION, T. E. P. A. T. C. O. T. E. (ed.).

European Commission. (2011). *Implementation of Directive 2009/31/EC on the geological storage of Carbon Dioxide, Guidance Document 4, Article 19 Financial Security and Article 20 Financial Mechanism*.

European Commission. (2015a). *Decision concerning the establishment and operation of a market stability reserve for the Union greenhouse gas emission trading scheme and amending Directive 2003/87/EU*.

European Commission. (2015b). *eu ets handbook*. european commission.

European Commission. (2017). *Funding innovative low-carbon technologies: the NER 300 program* [Online]. Available: https://ec.europa.eu/clima/sites/clima/files/docs/ner300_factsheet_en.pdf. Accessed 12 October 2019.

European Commission. (2018). *Directive (EU) 2018/410 of the European Parliament and of the council of 14 March 2018 amending Directive 2003/87/EC to enhance cost-effective emission reductions and low-carbon investments, and Decision* (EU)2015/1814 Brussels.

EYGM. (2014). *Understanding China's emissions trading schemes and emissions reporting*. EYGM Limited.

Fang, C. C. (2018). Carbon pricing, correcting climate change's market failure. *Sustainability: The Journal of Recording, 11*.

ICAP. (2019a). *ETS Fact sheets* [Online]. International Carbon Action Partnership. Available: https://icapcarbonaction.com/en/ets-map. Accessed 28 July 2019.

ICAP. (2019b). *China National ETS.*

IETA. (2014). *Use of offset credits across emission trading systems and carbon pricing mechanisms. IETA-Climate Challenges, Market Solutions.*

International Energy Agency. (2017). *Energy Technology Perspectives 2017.*

IPCC. (2018). *Special report on global warming of 1.5 °C (SR15).* Intergovernmental Panel on Climate Change.

Irlam, L. (2017). *Global costs of carbon capture and storage: 2017 update.* Global CCS Institute.

Liang, X., Lin, Q., Muslemani, H., Lei, M., Liu, Q., Li, J., Wu, A., Liu, M., & Ascui, F. (2019). Assessing the economics of CO_2 capture in China's iron/steel sector: A case study. *Energy Procedia, 158*, 3715–3722.

Mccormck, M. (2012). *A greenhouse gas accounting framework for carbon capture and storage projects.* Center For Climate and Energy Solutions.

MEE. (2019). *Interim regulations for the administration of carbon emission permit trading* (Consultation Paper). fgs.mee.gov.cn. Ministry of Ecology and Environment.

MOST. (2013). *The 12th five-year plan for national development of carbon capture, utilization and storage technology.*

Muslemani, H., Jiang, M., Chen, X. & Jia, Z. (2019). *Lower carbon technology approaches for steel manufacturing in China.* Working Paper 4.7 for the BHP Industrial CCS Project 'Unlocking the Potential of CCUS for Steel Production in China'. Edinburgh, UK.

Slater, H., Boer, D. D., Shu, W. & Guoqiang, Q. (2018). *The 2018 China carbon pricing survey.* Beijing.

NCEPU. (2019). *Techno-economic analysis of CO_2 capture technologies for the Iron/Steel industry in China.* Working Paper 4.11 for the BHP Iron and Steel CCS Project. North China Electric Power University, UK-China (Guangdong) CCUS Centre and The University of Edinburgh.

NDRC. (2013a). *Notice on greenhouse gas emission accounting methods and reporting guidelines for 10 sectors* (Trial).

NDRC. (2013b). *Notice on promoting carbon capture, utilization and storage demonstration.*

NDRC. (2017a). *NDRC Announcement, 2017 No.2* [Online]. National Development and Reform Commission. Available http://www.ndrc.gov.cn/zcfb/zcfbgg/201703/t20170317_841211.html. Accessed 11 October 2019.

NDRC. (2017b). *Notice on work plan for carbon emissions reporting, verification and monitoring for 2016 and 2017.* National Development and Reform Commission.

NDRC. (2017c). *Work plan for construction of the national emissions trading system* (Power Sector). National Development and Reform Commission.

NDRC. (2018). *Action plan on the establishment of a market-oriented and diversified ecological protection compensation mechanism.*

Pang, T., & Duan, M. (2016). Cap setting and allowance allocation in China's emissions trading pilot programs: Special issues and innovative solutions. *Climate Policy, 16*, 815–835.

State Council. (2006). *National mid- to long-term science and technology development plan (2006–2020).*

Stavins, R. N. (2001). *Experience with market-based environmental policy instruments.* Resources for the Future.

Tunnicliffe, H. (2017). *CCS: Why bother?* [Online]. The Chemical Engineer. Available: https://www.thechemicalengineer.com/features/ccs-why-bother/. Accessed 15 December 2018.

UNFCCC. (1998). *Kyoto protocol to the United Nations framework convention on climate change*. United Nations.UNITED NATIONS 2015. The Paris Agreement. Paris: United Nations.

Wang, X., Teng, F., Wang, G., Zhou, S., & Cai, B. (2018). Carbon leakage scrutiny in ETS and non-ETS industrial sectors in China. *Resources, Conservation and Recycling, 129*, 424–431.

World Bank Group 2018. (2019). State and trends of carbon pricing 2018. World Bank Group. *Carbon pricing dashboard* [Online].World Bank Group Available: https://carbonpricingdashboard.worldbank.org/map_data [Accessed 28 July 2019].

World Resources Institute & World Business Council for Sustainable Development. (2005). *The GHG protocol for project accounting*. USA.

World Steel Association. (2019). *Global crude steel output increases by 4.6% in 2018*[Online]. World Steel Association. Available https://www.worldsteel.org/media-centre/press-releases/2019/Global-crude-steel-output-increases-by-4.6--in-2018.html. Accessed 28 July 2019.

Xiong, L., Shen, B., Qi, S., & Price, L. (2015). Assessment of allowance mechanism in China's carbon trading pilots. *Energy Procedia, 75*, 2510–2515.

Xu, W., Wan, B., Zhu, T., & Shao, M. (2016). CO_2 emissions from China's iron and steel industry. *Journal of Cleaner Production, 139*, 1504–1511.

Zaman, P., & Yang, K. 2019. *Update on the Chinese National Emissions Trading Scheme-some progress but more questions than answers* [Online]. Reed Smith Client Alerts. Accessed October 2019.

Part IV
Projection

Achieving Carbon Neutrality by 2060: What (More) Has to Be Done?

7

Jun Fu, Jintao Xu, Yu Liu, Lunyu Xie, and Shilei Liu

On September 22, 2020, at the General Debate of the 75th Session of the UN General Assembly, Chinese President Xi Jinping delivered a speech and promised that China would peak carbon emissions prior to 2030 and try to achieve carbon neutrality by 2060.[1] That will be a very daunting task for China. In the wake of President Xi's speech, a workshop was held at the National School of Development at Peking University. The purpose was to look ahead and scope up what has to be done for China years down the road to achieve that daunting task. The authors of the present chapter all attended the workshop, and we agreed that there were essentially four approaches to the challenge, i.e., public policy approach (PPA), market-based approach (MBA), carbon capture and storage (CCS), and nature-based solutions (NBS). As the theme of climate change has much to do

[1] Subsequently the Chinese government redefied the goal of "achieving carbon neutrality by 2060" as including not just CO_2 but all GHGs.

The original version of this chapter was revised: A new sentence has been added on pages 254 and 272, respectively, in Sects. 2.1 and 4.4, and the symbol in the Table 2 note has been changed from "\$" to "¥" in Sect. 4.1. The correction to this chapter can be found at https://doi.org/10.1007/978-981-16-4310-1_8.

J. Fu (✉) · J. Xu · S. Liu
National School of Development, Peking University, Beijing, China
E-Mail: fujun@pku.edu.cn

Y. Liu
Institutes of Science and Development, Chinese Academy of Sciences, Beijing, China

L. Xie
Renmin University of China, Beijing, China

J. Fu et al. (eds.), *Climate Mitigation and Adaptation in China*,
https://doi.org/10.1007/978-981-16-4310-1_7

with the so-called "tragedy of the commons" (Hardin, 1968) on a global scale,[2] the role of PPA in connection with international cooperation and coordination has to be elevated.

Now armed with the conceptual framework of PPA + MBA + CCS + NBS (or PMCN model for short), our job was to build the model, get the data, and project forward with different scenarios such that we could help policy makers understand likely future outcomes as well as the implications of alternative policies. Intellectually, the dynamical PMCN modelling was both a deductive and inductive exercise within three sets of interrelated parameters, i.e., climate change, energy structure, and economic growth—along similar lines of the DICE model (dynamic integrated model of climate and the economy) developed by Nobel Laureate in economics William Nordhaus (Nordhaus and Boyer, 2000).

This chapter represents the results of that exercise. The chapter is organized into 4 sections. Section 1 puts China's climate mitigation and adaption efforts in global perspective. Section 2 elaborates on the framework of PPA + MBA + CCS + NBA in the Chinese context. Section 3 explains the mechanics of our model construction and scenarios planning. Projection results and key takeaways are presented in Section 4.

1 Section 1

1.1 Climate Change Mitigation and Adaption in Global Perspective

Climate change has posed a grave threat to mankind. Reduction of carbon emissions has increasingly become a global consensus. According to the IPCC (2018), in order to avert the worst impact of climate change on the eco-system on Earth, it is critical to limit global warming to 1.5 °C. This, in turn, requires a 45% reduction in global CO_2 emissions from 2010 level by 2030, and net zero CO_2 emissions world-wide by the middle of this century. Under the principle "common but differentiated responsibilities" of UNFCCC, while the contracting states to the Paris Agreement have been trying hard to reduce greenhouse gas emissions (GHG),[3] much more must be done. That is, more ambitious nationally determined

[2] Globally, the total amount of CO_2 emissions reached 40.1 billion tons in 2019, about 86% of which was from fossil-fuels and 14% from land use. Of the total, 31% was sequestered by the land, 23% by the ocean, and the rest (or 46%) went up into the atmosphere.

[3] In the composition of GHG defined Kyoto Protocol, Carbon Dioxide (CO_2) takes about 81%, Methane (CH_4) 10%, Nitrous Oxide (N_2O) 7%, and the rest are Fluorinated Gases (e.g.,

contributions (NDC) must be adopted to make sure that the goal of staying within 1.5 °C will be achieved. NDCs are at the heart of the Paris Agreement to reduce CO_2 missions. They embody voluntary efforts by each contracting state to reduce CO_2 emissions and adapt to the impact of climate change.

In November 2018, the UN released a special report—i.e., the IPCC's 1.5 °C Special Report—urging the contracting states to the Paris Agreement to do more. In light of this, many jurisdictions across the world—including China, the European Union (EU), the United Kingdom (UK), Canada, South Korea, New Zealand, New York, Hawaii and California—renew efforts and set climate targets to carbon neutrality. Carbon neutrality refers to the target to remove as much carbon dioxides from the atmosphere as they are put into it. The overall goal of carbon neutrality is to achieve a zero carbon footprint. In December 2019, the UN announced that 73 countries, 14 regions and 398 cities around the world had started to work towards carbon neutrality by 2050 (UNFCCC, 2019).

Based on the Paris Agreement in 2015, the European Union (EU) had committed itself to the binding target of at least a 40% reduction in GHG emissions from the 1990 level. Having cut a so-called "European Green Deal", the EU has aimed higher—to achieve a carbon neutral goal by 2050. Specifically, the green deal has raised the EU targets to at least 50% reduction and has started to move towards 55% GHG emissions reduction by 2030. And in line with the EU goal of reaching carbon neutrality by 2050, many European countries have adopted NDCs toward zero net GHG emissions. Austria, Denmark, France, Germany, and Sweden have all put climate policies in place, aiming at carbon neutrality by 2040 (Austria), by 2045 (Sweden), and by 2050 (the rest) respectively.

In the United Kingdom (UK), UK Climate-change Act entered into force in June 2019. The Act has confirmed that the UK will achieve zero net carbon emissions by 2050. Under the UK Climate-change Act, Scotland aims to achieve carbon neutrality in 2045, and Welsh will reduce CO_2 emission by 90% from 2010 levels by 2030. Similarly, New Zealand amended its Climate Change Response Act in 2019. The Amended (Zero Carbon) Act aims at net zero GHG emissions by 2050. As the agricultural sector accounts for a large part of New Zealand's GHG emissions, the Zero Carbon Act also aims to reduce biogenic methane emissions by 24–47% from 2017 levels by 2050.

As a developing country in the midst of rapid industrialization and urbanization, China has been working hard to catch up in terms of decarbonization. In 2009, China had announced the goal of a 40%-45% reduction in carbon emissions

---

hydrofluorocarbon (HFCs), perfluorocarbon (PFCs), and sulfur hexafluoride (SF_6)), about 3%.

per GDP by 2020, as compared to the 2005 level. In 2015, China set its NDCs to cap CO_2 emissions no later than 2030. According to China's Climate Change Policy and Action 2019 Annual Report,[4] carbon emissions per GDP had dropped by 45.6% from 2005 level by 2018, exceeding the original reduction target.

As the world's biggest developing country and CO_2 emitter, China's new voluntary pledge—i.e., to achieve carbon neutrality in 2060—would contribute the lion's share of the reductions of CO_2 emissions world-wide. In comparative perspective, the deadline that China sets for itself to achieve carbon neutrality is very tight. Whereas the EU has planned to spend 70 years from carbon emissions peak to carbon neutrality, and the United States 45 years, China only gives itself a 30-year transition period. The task is daunting. To illustrate, in 2019, fossil fuels still accounted for about 85% of China's energy structure, of which coal was 58%, the country's carbon emission per capital remained 46% higher than the world average; and total CO_2 emissions about 29% of the world's total. For China to peak carbon emissions prior to 2030 and achieve carbon neutrality in 2060, it must move aggressively to de-carbonize[5] its economy in the year ahead—indeed much more aggressively than the developed countries.

(a) What are the available strategies of carbon emission reduction in China to achieve carbon neutrality by 2060?
(b) What are the effects of different mitigation strategies on carbon emission, energy demand and macro economy?
(c) Whether can China achieve carbon neutrality by 2060 through the current available strategies and at what cost?

To try to answer the above questions systematically, we have developed a conceptual framework of PPA + MBA + CCS + NBS, so as to delineate the parameters of our model building.

[4] The Ministry of Environmental Protection from 2009 publishes the China's Climate Change Policy and Action Annual Report annually.

[5] Carbon removal or decarbonization is a term that includes many forms of carbon removal from the atmosphere, whether through land sequestration or through negative emissions technologies that actively pull CO_2 out of the atmosphere, such as direct air capture or biomass energy with carbon capture and storage.

2 Section 2

2.1 A Conceptual Framework: PPA + MBA + CCS + NBS

Concept is critical in generating knowledge from experience. Or the world would remain too chaotic for us to understand. The fewer fundamental concepts that lies at the base of a science, the higher the level knowledge to which it raises us. Add actual context, we practice in the real world. Call it *theoria cum praxi* philosophically. There are essentially four approaches regarding mitigation and adaptation to the impact of climate change. They are: public policy approach (PPA), market-based approach (MBA), carbon capture and storage (CCS), and nature-based solutions (NBS). Let us elaborate a bit more on each of them, starting backwards from NBS.

Nature-Based Solution (NBS)

The International Union for the Conservation of Nature (IUCN) defines nature-based solution (NBS) as actions to protect, sustainably manage, and restore natural or modified ecosystems that support the achievement of society's development goals and address major societal challenges, such as climate change, food security, water security, human health, disaster risk, etc.[6]

In climate change, NBS has been recognized as an important measure to help and protect human beings from the impact of climate change while slowing down further warming (Cohen-Shacham et al., 2016; Nature, 2017). Forests, as the main resource of NBS, sequester CO_2 as trees grow and have potential to mitigate the impact of climate change (Bellassen & Luyssaert, 2014). Xu et al. (2020) provides the estimate of China's forest carbon sequestration potential. Based on their estimation, by the end of 2050, the carbon sinks of the existing forest in 2050 will be 1.54 billion tons/year. He et al. (2020)'s prediction is much lower than Xu et al. (2020)'s estimation. According to their research, China's forest carbon sequestration will be 700 million tons/year by 2050. The reason for the difference is the emission reduction effect of wood products does not enter into He et al. (2020)'s prediction.

CO_2 Capture and Storage (CCS)

CO_2 capture and storage (CCS), as our previous chapters have explained, refers to the process that carbon dioxide is separated from industrial or related emission sources, transported to storage sites, and isolated from the atmosphere. In recent years, China has ramped up the R&D of CCUS (carbon capture, utilization and storage) through the National Basic Research Program of China (aka.,

[6] https://www.iucn.org/commissions/commission-ecosystem-management/our-work/nature-based-solutions.

972 Program), the National Hi-Tech Research and Development Program of China (aka., 863 Program), and National Science and Technology Support Program. One illustration of the "U" part (CO_2 utilization) is CO_2-EOR (enhanced oil recovery) technology, which, if successful, could result in negative CO_2 emissions. It injects CO_2 into an underground geologic zone that contains hydrocarbons to produce more oil. So far there are about 30 CCS/CCUS pilot or demonstration projects either already in operation or under construction in China, and mostly in oil fields and coal-fired power plants.

CCS is not only a large-scale CO_2 emission reduction technology with great potentials, but also the only technology available that can deliver large-scale direct emission reductions for energy-intensive industries, and with minimal restructuring costs on existing energy structure. Bioenergy with CO_2 capture and storage (BECCS) is the installation of CCS technology in the biological processing industry or biofuel power plants. Both CCS and BECCS create negative carbon emissions. As He et al. (2020) estimate that the carbon sink of CCS and BECCS technology in China would be 510 million tons and 880 million tons in 2050 under 1.5 °C and 2 °C scenarios, respectively. At present, however, the biggest hurdle is the prohibitively high cost for CCS, estimated at \$200–300 tCO_2. As 1 ton of coal would produce roughly 2 tons of CO_2, it means that it would cost well over \$400 to sequester CO_2 emissions from the burning of 1 ton of coal.

Market-Based Approach (MBA)

On way to understand how market-based approach (MBA) works to drive decarbonization is carbon markets, or known in Europe as emission trading systems (ETS). Broadly speaking, ETS can be understood as a financial way to internalize environmental benefits when investing in sustainable energy systems. ETS contributes to economic efficiency by facilitating CO_2 emission reductions where it is cheapest to achieve them. Polluters or emitters who would find it costly to reduce their emissions are allowed to buy emission allowances from polluters or emitters who can abate at lower costs. In a "perfectly" market, the costs associated with reducing an additional unit of emissions would be equalized, and total costs of reaching a given environmental or climate target would be minimized. As discussed in the previous chapters, China started 7 regional ETS pilots in 2011 (in Beijing, Chongqing, Guangdong, Shanghai, Shenzhen, Tianjin, and Hubei), and a national ETS in 2017, although the national ETS covers only the power sector so far but is expected to expand its coverage years down the road.

Another example of MBA is green bonds. Green bonds can enable fund-raising and investment for both existing and new projects with environmental benefits, e.g., CCS technology presumably in the years ahead. Market principles of information disclosure and transparency are critical in developing green bond markets. They help potential investors by ensuring availability of information necessary to evaluate the environmental impact of their investments; and they also assist underwriters by moving the market towards standard disclosures which can facilitate transactions. In December 2015, China's green bond market was launched, when the People's Bank of China (PBOC) first came out with a catalogue of green projects and introduced green bonds in the interbank bond market. Subsequently, a total of 9 market pilots were rolled out for China to test the ground and learn from experience of green finance, including green loans, green bonds, and green insurance. At represent, PBOC is considering upgrading the green bond catalogue by removing fossil fuel projects and working with international counterparts to align green taxonomies.

Public Policy Approach (PPA)

An extensive literature exists, both theoretical and empirical, that supports the argument that when externalities exist, public policy becomes powerful instruments to correct market failures. Combating externalities—of which CO_2 emissions, water pollutions and environmental degradations are textbook examples—is a legitimate arena for government intervention using what we call public policy approach (PPA). The argument is all more compelling when the ecosystem on Earth is at stake. To deal with externalities or market failures, PPA instruments essentially include (a) a Tax, (b) a Subsidy, (c) a Restriction, Standard, Target or Quota.

Experts have identified a number of policy instruments that are efficacious in restricting or reducing carbon emission (Gallagher et al., 2019). In addition to national emission-trading system (ETS) and green bonds, which we have discussed under market-based approach (MBA),[7] the policy instruments include:

- Subsidies for renewable energy (feed-in tariffs).
- Energy efficiency policies for power plants.
- Non-fossil targets
- Coal cap policies

[7] Note that at the fundamental level this categorization is debatable. For example, in the scheme of "cap and trade" of CO_2 emissions in ETS, the "cap" is ultimately set by the government and then the firm get to "trade". And arguably, to the extent that ETSs in China are still in the piloting stage and carbon pricing is still subject to government guidance, they can be discussed generally under PPA as well. More pointedly, in a transitional economy like China's, whether or not to set up these markets can be understood as a public policy choice.

- Energy efficiency standards for buildings and equipment.
- Key enterprise programs.
- Economic structural reforms.
- Removing fossil fuel subsidies.
- Energy intensity targets
- Air pollution standards
- Differentiated electricity tariffs.
- Phasing out old and inefficient power plants.
- Electricity sector reform.
- Innovation policies
- Environmental resource tax

Note that when we reviewed China's evolving regulatory framework for a low-carbon economy in Chapters 1 and 2 of this study, we found that the variants of these public policy tools listed above have all existed in China. For the purposes of the current study, we put these policy instruments under PPA, and sum them up into three major categories of strategies, i.e., energy efficiency enhancement, cost reduction in energy renewables, and market inclusion and carbon pricing. Below, we elaborate a bit more on them one by one.

Energy Efficiency Enhancement

Generally, carbon emission reduction is a low-carbon strategy, and can be achieved through improving efficient use of fossil energy and reducing the energy consumption of industrial sectors. This appears to be the predominant strategy China has deployed to achieve CO_2 peak before 2030, and there is good reason. According to the statistics from World Bank, the energy intensity of primary energy in China is almost twice as much as German by 2015 (Fig. 1), which indicates China has huge potential to lower the energy intensity and increase utilization efficiency. Ding et al. (2019) calculate the energy intensity of China from 2000 to 2100 based on the change trend of energy intensity from 1980 to 2006. They predict that the annual energy intensity in China would decrease by 3.92% per year. Zhang et al. (2019) forecast the energy intensity for 41 industries of China from 2020 to 2030. They find that the energy intensity falls by 16.1% under scenario of no policy guidance and by 31.9% under scenario with policy guidance. Song et al. (2018) and Niu et al. (2016) make predictions about the energy intensity in China, too. According to above predictions, China's future energy intensity change located in the range of 1% to 4%. The improvement in energy efficiency has played and will continue to play an important role in helping China peak CO_2 emissions prior to 2030 and achieve carbon neutral in 2060.

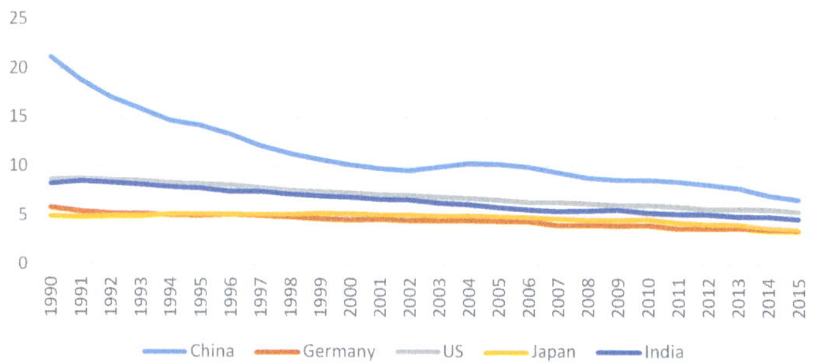

Fig. 1 Energy intensity: China, Germany, US, Japan and India (1990–2015). *Source* World Bank. Unit: MJ/$2011 PPP GDP

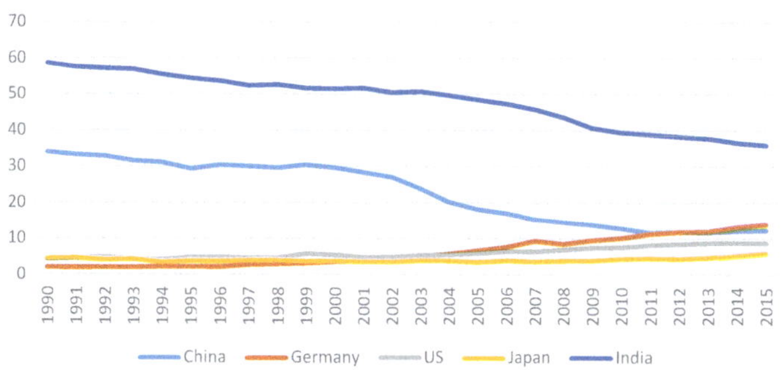

Fig. 2 Share of Renewable Energy: China, Germany, US, Japan and India (1990–2015). *Source* World Bank Unit: %

Cost Reduction in Renewable Energy

Ultimately, renewable energy is a technological path toward zero-carbon emissions. If a good strategy is a sequence of interconnected foci, then the strategic focus of zero-carbon emissions will have to be in place after China peaks CO_2 emissions by 2030. In the meantime, the decline in the cost of renewable energy such as wind and solar energy as well as biomass energy will promote the transformation toward clean energy and achieve carbon emission reduction. China is experiencing a transition period from a firewood-dominated old era of renewable energies toward a wind energy-, solar energy- and biomass energy- dominated new era of renewable energies (Fig. 2).[8] Based on the estimation of He et al. (2020), the average cost for centralized photovoltaic solar panels, onshore wind energy and battery storage dropped by 77%, 35% and 85% respectively from 2010–2018. Li et al (2020) finds that the cost of photovoltaic power generation has decreased from 100$/kwh in 1990 to 0.05$/kwh in 2020 and the annual decrease is as high as 22.38%. Although China has strived to increase investment in R&D of renewables, the share of renewable energy in China is still below 20%. The aim of carbon neutral by 2060 will require a much lower renewable energy cost and a much higher the proportion of the use of renewable energy.

Market Inclusion and Carbon Pricing

It is being widely adopted to promote carbon emission reduction by pricing carbon emission, such as imposing carbon taxes on industrial sectors and residents, establishing carbon markets (e.g. California' cap and trade, EU's emission-trading schemes or ETS). In 2011, China announced to launch seven regional ETS pilots in two provinces (Hubei and Guangdong) and five cities (Beijing, Shanghai, Shenzhen, Tianjin and Chongqing). These regional ETS pilots are intended to gain experience associated with running carbon markets and to identify problems that should be resolved before moving to a nationwide carbon market. There have been many researches predicting the future carbon price in China.

In 2017, China rolled out a national carbon market but it only included the power sector. It is expected that over time the national market will also include other industrial sectors such as steel and cement, each contributing about 15% to the total industrial CO_2 emissions in China, respectively. Yuan et al. (2020) estimates that the carbon price will reach the highest (106.5$/t$CO_2$) by 2090 with reference to the future carbon price calculation formula of Wilkerson et al. (2015). The scenario given by Dong et al. (2017) show that the highest carbon price will reach 120$/t$CO_2$. Above predictions are based on climate goal that China's carbon emission would peak no later than 2030. The aim of carbon neutral by 2060 urges for higher carbon price levels. This is especially the case, if China eventually connects its domestic ETSs with global ETSs.

[8] India is the highest in the proportion of renewable energy among the five countries in Fig. 2 as India is still dominated by firewood.

3 Section 3

3.1 Model Construction and Scenarios Planning

To operationalize our PMCN (PPA + MBA + CCS + NBS) model, we draw upon the tool of recursive dynamic CGE model representing the Chinese economy, named CHINAGEM,[9] and fuse it with the parameters of the PMCN model. Before we proceed further, a word of CHINAGEM is in order. CHINAGEM has been widely used in the evaluations of economic and environmental policies in China. This model is capable of predicting the effect of macro policies on key aggregate indexes, such as GDP, price index, trade and other macroeconomic indicators, as well as industry-level outputs and prices. It contains 159 economic sectors, six economic subjects (manufacturers, investment, households, government, exports, and inventories), and nine distribution sectors (wholesale & retail, warehousing, insurance, rail transport, road transport, water transport, air transport, pipeline transport, transport services).

Carbon emissions from fossil fuel combustion and industrial production processes (excluding energy combustion) are modelled in CHINAGEM. The coefficients of carbon emissions used to calculate the carbon emission from fossil energy combustion are from IPCC AR5 (2014). China's fossil energy emission from 2017 to 2020 is calibrated according to the IEA World Energy Outlook (2020). In the calculation of carbon dioxide emissions in the industrial production process, we mainly consider the emissions in the cement production, because it accounts for over 70% of the emissions from all the industrial production processes (excluding energy combustion). According to Shan et al. (2019), carbon emissions from cement production in China in 2017 were 7004.4 million tons.

Carbon emission and production input are considered as the two main inputs of cement industry. In the CHINAGEM model, the emission reduction is described as the partial substitution relationship between carbon emission and production input. When the carbon price increases, the cement sector will reduce carbon emission and use more production inputs. In economic sense, cement producers will adopt more carbon reduction technologies when facing higher carbon price, but they will also face higher emission reduction cost. According to Hyman et al. (2002), the substitution elasticity of carbon emission and production input is set as 0.11.

Carbon emission accounts in the CHINAGEM model are based on fossil energy combustion and industrial production within the whole economic system. The carbon emission accounts are used to predict future carbon emissions in China through 2060.

[9] CHINAGEM is developed by the policy research center of Victoria University, Australia.

3.2 Set-Up of Model Parameters

Based on the combination of the PMCN and CHINAGEM models, we analyze the strategic paths leading to net zero carbon emissions by 2060 in China, under several PPA scenarios, including carbon pricing. In setting up key parameters, we first construct the baseline scenario, covering China's GDP, population, labor force and industrial macroeconomic structure from 2020 to 2060, based on the forecasts of the World Bank, CEPII and other international institutions. Second, drawing upon best estimates by the current studies, we take NBS (e.g., carbon sinks such as forests) and CCS as predetermined, and focus on PPA for de-carbonization of industrial processes, either directly (e.g., deployment of (CCS) or indirectly (e.g., energy efficiency enhancement). According to the literature discussion in the previous section, we choose two carbon neutral targets for NBS and CCS—one conservative and one optimistic:

(1) Conservative target: by 2060, NBS and CCS reaches 1.55 billion tons/year (i.e., 700 million tons (NBS) + 750 million tons (CCS)), implying China's CO_2 emissions need to be reduced to 1.55 billion tons by NBS and CCS in 2060, so that the carbon emissions and carbon sinks can be basically offset, reaching net zero carbon emission;
(2) Optimistic target: by 2060, NBS and CCS reaches 2.8 billion tons/year (i.e., 1.6 billion tons (NBS) + 1.2 billion tons (CCS)), implying that China's CO_2 emissions need to be reduced to be 2.8 billion tons by NBS and CCS in 2060, so that the carbon emissions and carbon sinks can be basically offset, reaching net zero carbon emissions.

We subsume MBA (e.g., ETSs and green bonds) under PPA, because the future of these market-oriented pilots in China continues to be guided if not totally controlled by the government, although their expansions are expected to be sped up. Under PPA, the policy portfolios include three major categories, i.e., improvement in energy efficiency, cost reduction in renewable energy, and carbon pricing. In analyzing PPA efficacy, we consider the scenarios of single PPA and composite or comprehensive PPA.

3.3 Single PPA Scenarios

(1) *Improvement in energy efficiency*

As the existing studies forecast that China's improvement in energy efficiency will be concentrated in the range of 1% - 4%, we assume the overall energy intensity of industrial sectors decline by the average rate of 0.5%, 1%, 1.5%, 2%, 2.5% and 3%[10] respectively and we simulate the corresponding policy scenarios.

(2) *Cost reduction in renewable energy in China*

For the cost reduction in renewable energy, we rely on the forecast of the International Energy Agency (IEA) on the future cost of photovoltaic and wind power generation. Considering that the power generation cost of renewable energy such as photovoltaic has been at a low level and the decline rate will gradually narrow in the future, we set two scenarios of cost reduction in renewable energy in China: one with high rate and the other with low rate. In the high rate of cost reduction scenario, it is assumed that the costs of solar photovoltaic, onshore wind power and offshore wind power decrease by 3.25%, 1.06% and 4.27% respectively each year in 2020–2060. In the low rate of cost reduction scenario, the costs decrease by 1.62%, 0.53% and 2.13% respectively. At the same time, in the CHINAGEM model it is assumed that the government surcharges on fossil energy to promote the reduction of renewable energy cost.

Referring to the method of Dong et al. (2017), this study sets six scenarios of carbon pricing in which the carbon price in 2060 are 100, 200, 300, 400, 500 and 600\$/tCO_2 respectively and assumes that the carbon price increases exponentially. At the same time, it is assumed that the carbon market will have included all major industrial sectors such as power, cement and steel, and that government will transfer all carbon emission tax revenue of carbon pricing policy to residents. The various single PPA scenarios are summarized in Table 1.

3.4 Composite PPA Scenarios

Scholars have proposed that composite or complex PPA portfolios give better results than single PPA for emission reduction. It is believed under composite or complex PPA portfolios the negative impact on the economy will be significantly reduced. For example, Song et al. (2018) find the policy combination can accelerate the peak time of carbon emission, so that China can reach the

[10] To be conservative, we do not select 3.5% and 4% as the decline rate.

Table 1 Single PPA scenarios

Strategy	Abbr		Shock	Literature
Improvement in energy efficiency	E	E1	0.5%	Ding et al. (2019) Zhang et al. (2019) Yuan et al. (2014) Song et al. (2018) Niu et al. (2016)
		E2	1%	
		E3	1.5%	
		E4	2%	
		E5	2.5%	
		E6	3%	
Cost reduction in renewable energy	R	R1	high rate	He et al. (2020) Li (2020) IEA (2020)
		R2	low rate	
Carbon pricing	T	T1	100$/t CO_2	Duan et al. (2018) Yuan et al. (2020) Wilkerson et al. (2015) Dong et al. (2017)
		T2	200$/t CO_2	
		T3	300$/t CO_2	
		T4	400$/t CO_2	
		T5	500$/t CO_2	
		T6	600$/t CO_2	

peak in 2026 when the annual GDP growth rate is not less than 5%. Lin and Jia (2019) also proposed that the simultaneous implementation of carbon trading, energy efficiency improvement, carbon sink and other emission reduction measures would help China reach the peak by 2030. Here, in a way that ranges from the optimistic approach of William Nordhaus (2007) to the pessimistic approach of Nicholas Sterns (2006) regarding global warming,[11] we test 72 PPA combinations of improvement in energy efficiency, cost reduction in renewable energy and carbon pricing, as shown in Table 2.

[11] Generally, Nordhaus assumes a high discount rate between the present and the future and therefore proposes a more gradual carbon tax than Stern, who assumes a lower discount rate and advocates stronger and immediate action.

Table 2 Composite PPA Scenarios

Scenarios	R	A	T	Scenarios	R	A	T
1	R1	A6	T6	37	R2	A6	T6
2	R1	A6	T5	38	R2	A6	T5
3	R1	A6	T4	39	R2	A6	T4
4	R1	A6	T3	40	R2	A6	T3
5	R1	A6	T2	41	R2	A6	T2
6	R1	A6	T1	42	R2	A6	T1
7	R1	A5	T6	43	R2	A5	T6
8	R1	A5	T5	44	R2	A5	T5
9	R1	A5	T4	45	R2	A5	T4
10	R1	A5	T3	46	R2	A5	T3
11	R1	A5	T2	47	R2	A5	T2
12	R1	A5	T1	48	R2	A5	T1
13	R1	A4	T6	49	R2	A4	T6
14	R1	A4	T5	50	R2	A4	T5
15	R1	A4	T4	51	R2	A4	T4
16	R1	A4	T3	52	R2	A4	T3
17	R1	A4	T2	53	R2	A4	T2
18	R1	A4	T1	54	R2	A4	T1
19	R1	A3	T6	55	R2	A3	T6
20	R1	A3	T5	56	R2	A3	T5
21	R1	A3	T4	57	R2	A3	T4
22	R1	A3	T3	58	R2	A3	T3
23	R1	A3	T2	59	R2	A3	T2
24	R1	A3	T1	60	R2	A3	T1
25	R1	A2	T6	61	R2	A2	T6
26	R1	A2	T5	62	R2	A2	T5
27	R1	A2	T4	63	R2	A2	T4
28	R1	A2	T3	64	R2	A2	T3
29	R1	A2	T2	65	R2	A2	T2
30	R1	A2	T1	66	R2	A2	T1

(continued)

Table 2 (continued)

Scenarios	R	A	T	Scenarios	R	A	T
31	R1	A1	T6	67	R2	A1	T6
32	R1	A1	T5	68	R2	A1	T5
33	R1	A1	T4	69	R2	A1	T4
34	R1	A1	T3	70	R2	A1	T3
35	R1	A1	T2	71	R2	A1	T2
36	R1	A1	T1	72	R2	A1	T1

Note Renewable energy policy: R1 cost of renewable energy decreases rapidly, R2 cost of renewable energy decreases slowly; energy efficiency policy: A1 energy efficiency increases by 0.5% /year, A2 energy efficiency increases by 1%/year, A3 energy efficiency increases by 1.5%/year, A4 energy efficiency increases by 2%/year, A5 energy efficiency increases by 2.5%/year, A6 energy efficiency increases by 3%/year; carbon price policy: T1 ¥100/tCO_2 in 2060, T2 ¥200/tCO_2 in 2060, T3 ¥300/tCO_2 in 2060, T4 ¥400/tCO_2 in 2060, T5 ¥500/tCO_2 in 2060, and T6 ¥600/tCO_2 in 2060

4 Section 4

4.1 Projection Results Under Different Scenarios

Under the baseline scenario, China's total carbon emissions in 2060 are 8.812 billion tons, including 7.746 billion tons from fossil energy combustion.

(1) Improvement in energy efficiency

Under single PPA scenarios—energy efficiency improvement, the energy use efficiency of the industry and the residential sector has been improved, which eventually reduces the demand for all energy products. The output of fossil energy sector drops significantly, which leads to a significant reduction of carbon emissions compared to the baseline scenario.

The total carbon emissions are significantly reduced compared to the baseline scenario. By 2060, the total carbon emissions of the A1-A6 scenarios drop to 7.83 billion tons, 7.0 billion tons, 6.30 billion tons, 5.69 billion tons, 5.17 billion tons and 4.72 billion tons respectively (Fig. 3), which are reductions of 11.2–46.4%, as compared with the baseline scenario.

Carbon emissions may peak earlier than 2028, but they are still far from achieving the carbon neutral target in 2060 (Fig. 4).

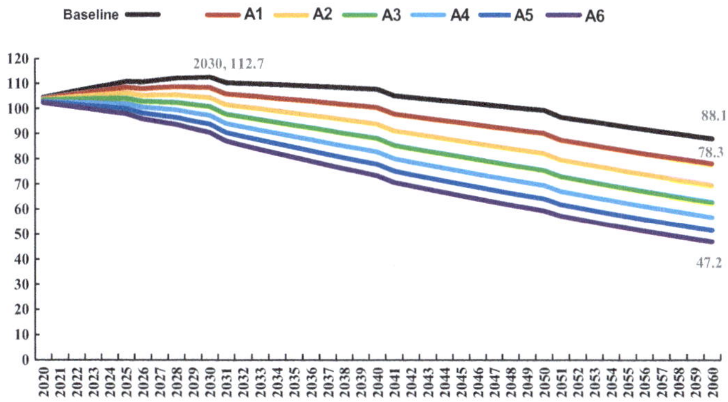

Fig. 3 CO_2 emissions in China (2020–2060) by various scenarios of energy efficiency improvement (Unit: 0.1 billion tons) Note: A1 energy efficiency increase by 0.5%/year, A2 energy efficiency increase by 1%/year, A3 energy efficiency increase by 1.5%/year, A4 energy efficiency increase by 2%/year, A5 energy efficiency increase by 2.5%/year, A6 energy efficiency increase by 3%/year

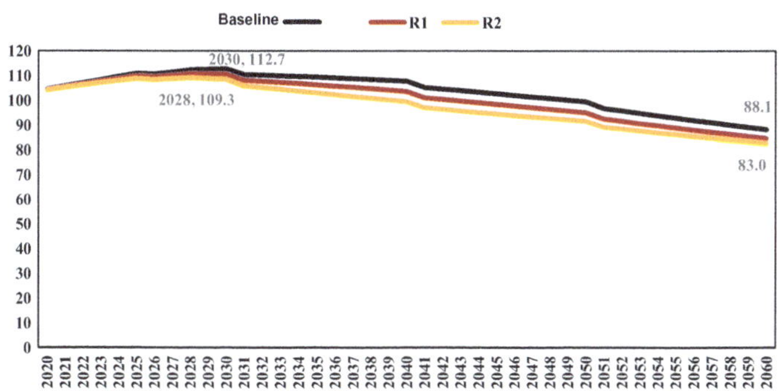

Fig. 4 CO_2 Emissions in China (2020–2060) by various scenarios of renewable energy cost reduction (Unit: 0.1 billion tons) Note: R1 Cost of renewable energy decreases rapidly, R2 Cost of renewable energy decreases slowly

(2) *Cost reduction in renewable energy*

In renewable energy cost reduction scenarios, the cost reduction of wind power, solar power and biomass power will reduce the average price of power, leading to the substitution of coal power and gas power and the reduction of fossil energy consumption. As shown in Fig. 4, by 2060, the emissions of low rate cost reduction (R1) and high rate cost reduction (R2) scenarios are 8.48 billion tons and 8.3 billion tons, respectively, which are reductions of 3.8% and 5.8% compared with the baseline. The emission reduction effect is weaker than that of the energy efficiency-improving scenarios, which is far from the carbon neutral target.

(3) *Market inclusion and carbon pricing*

In the scenario of increasing carbon tax, carbon emissions decrease to 7.93 billion tons, 7.51 billion tons, 7.19 billion tons, 6.94 billion tons, 6.73 billion tons and 6.55 billion tons in 2060, respectively (Fig. 5). Compared with the baseline in 2060, the emission reduction ranges from 880 million tons to 2.26 billion tons, which are reductions of 10.0–25.7%. The total emission differences among different carbon tax scenarios showed an increasing trend over time, possibly because that we set carbon tax to increase exponentially.

(4) *Composite PPA scenarios*

As shown in Fig. 6, the composite PPA effectively reduces China's total carbon emissions from 2020 to 2060, but there is still a certain gap with the carbon neutral target in 2060. As previously mentioned, in 2060, China's negative emissions

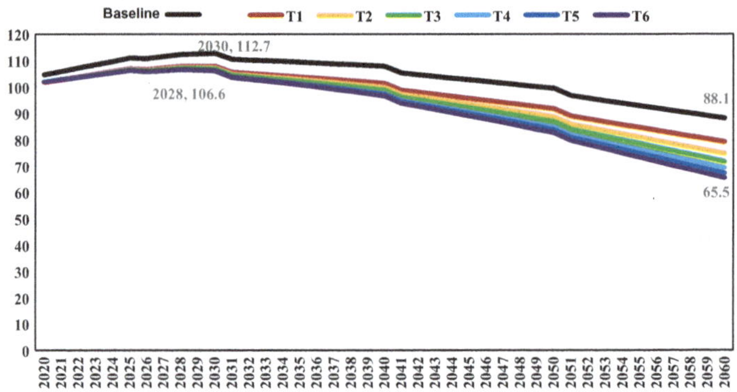

Fig. 5 CO_2 Emissions in China (2020–2060) by various scenarios of carbon pricing (Unit: 0.1 billion tons). *Note* T1 $100/t$CO_2$ in 2060, T2 $200/t$CO_2$ in 2060, T3 $300/t$CO_2$ in 2060, T4 $400/t$CO_2$ in 2060, T5 $500/t$CO_2$ in 2060, and T6 $600/t$CO_2$ in 2060

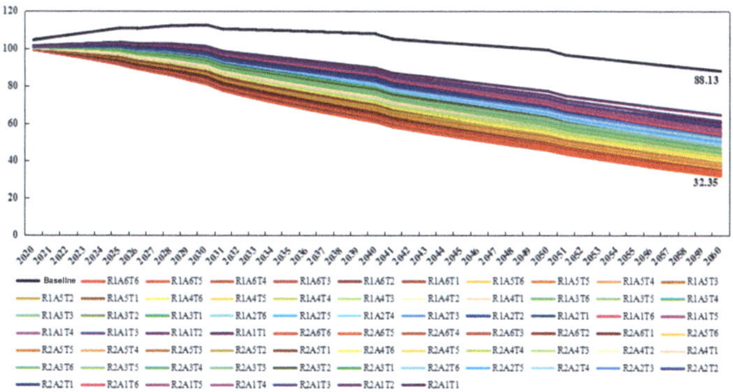

Fig. 6 CO_2 Emissions in China (2020–2060) by various scenarios of policy portfolios (Unit: 0.1 billion tons). *Note* As an example, R1A6T6 indicates a combination of policy R1, A6 and T6

from forest carbon sinks and CCS technologies will not exceed 3 billion tons. Therefore, the carbon neutral requires that the total carbon emissions in 2060 should be controlled within 3 billion tons. Under the most active combination of PPA instruments for emission reduction (R1T6A6), the total carbon emission in 2060 is 3.235 billion tons, which is 235 million tons higher than the 3 billion tons' target (Fig. 6). As a conclusion, only relying on carbon price, energy efficiency and renewable energy policy instruments cannot meet the requirements of the 2060 carbon neutral target.

4.2 Impact on Energy Structure

Composite PPA rather than single PPA for emission reduction reshapes the future structure of energy demand more powerfully. In the baseline scenario, the energy demand in 2060 is 6.239 billion tons of standard coal, with coal accounted for 31.5%, renewable energy accounted for 35.9%, and oil, natural gas and nuclear energy accounted for 12.2%, 14.9% and 5.4% respectively (Fig. 7). Under composite PPA scenarios, the total energy demand in 2060 will decrease to 2.22–5.49 billion tons of standard coal. Renewable energy accounts for 27.3–77.5% of the total energy demand, while coal accounts for 8.1–41.6%, oil accounts for 11.4–13.9%, natural gas accounts for 2.0–17.7%, and nuclear energy accounts for less than 4%.

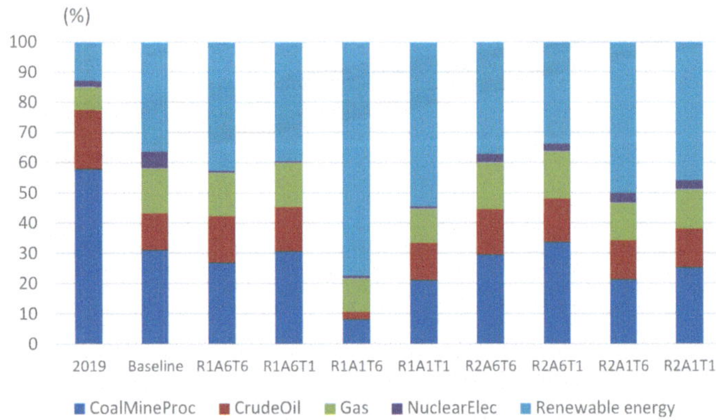

Fig. 7 Energy Structure in China in 2060 by various scenarios of policy portfolios (Unit: %)

4.3 Impact on Economy

Climate mitigation and adaptation comes at the cost of economic loss. We find that (1) GDP loss and emission reduction show an obvious positive correlation (Fig. 8). That is, to obtain greater carbon emission reduction, we need to bear greater GDP loss; (2) Compared with energy efficiency improvement and carbon price policies, the rapid cost reduction of renewable energy has higher emission reduction potential and smaller economic loss cost; (3) Based on the tradeoff

Fig. 8 Tradeoff between CO_2 Emission and GDP Loss in China in 2060, by various scenarios of policy portfolios *Note* Blue dot represents some one policy portfolio

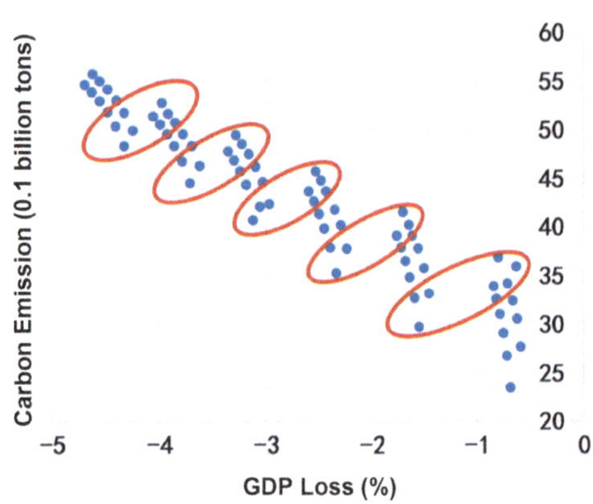

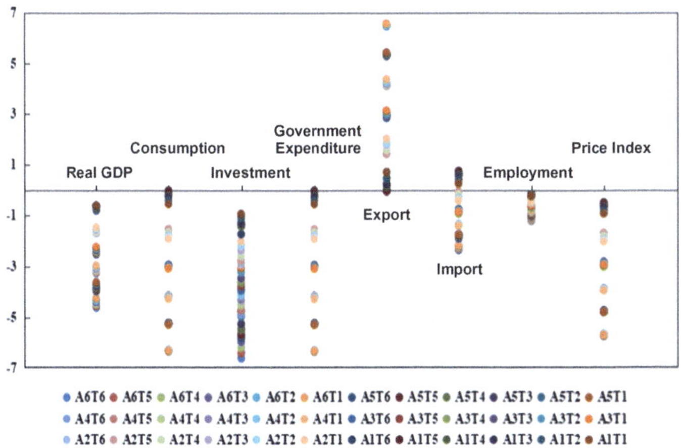

Fig. 9 Impact of policy portfolios on macroeconomic variables in China in 2060

between emission reduction benefits and economic loss, the policy combination of low energy efficiency improvement and high carbon price is better than that of high-energy efficiency improvement and low carbon price.

For other macroeconomic variables (Fig. 9), we find that: (1) A complex PPA strategy for emission reduction significantly decreases the output of fossil energy sector; (2) Energy efficiency policy not only improves energy efficiency, but also reduces labor demand of industrial sector, and therefore leads to a decrease in total employment by a magnitude of 0.13–1.22%; (3) A complex PPA strategy increases the use cost of fossil energy and reduces the demand for fossil energy products, and thus reduces the investment in the fossil energy sector; (4) The loss of economy causes the decline of national income, which makes the consumption of residents and government expenditure decrease in different degrees; (5) The decline of energy demand reduces the price of energy products; (6) The export demand increases in almost all complex PPA scenarios but the import demand is the opposite. The lower price level has a pulling effect on export. Meanwhile, the rise of energy cost weakens the competitiveness of non-energy products, making the import of products increase.

4.4 Concluding Remarks

Key Takeaways

To achieve carbon neutral goal by 2060, improvement in energy efficiency, cost reduction in renewable energy and carbon pricing are all necessary, but not enough even in a combination of all above strategies. New policy instruments, especially for a more powerful shift from fossil to no-fossil fuels, are needed to achieve the carbon neutral goal in time.

All the strategies of carbon emission reduction come at the cost of GDP loss, with modest declines in employment, investment, and consumption. Having said that, however, a better calibrated mix of policy instruments will take smaller costs of economic loss. In this study, we recommend the combinations of low energy efficiency improvement and high carbon price, including carbon taxation rather than campaign-style, across-the-board prohibition.[12] In other words, on balance, MBA is the preferred approach.

Combined PPA aiming at de-carbonization will reshape the future structure of energy demand dramatically. Renewable demand will increase and the fossil energy demand, especially coal demand, will decrease. Yet a caveat is in order here. Even in the most comprehensive of all PPA scenarios available today, by 2060, renewable energy will only accounts for less than 80% of the total energy consumption,[13] while fossil energy still accounts for more than 20%. Promoting electric vehicle sector development and other forms of electrification would become especially important for further reductions in fossil energy demand. In this context, as a matter of rational strategic sequencing, priorities should be given for the electricity sector to achieve zero carbon ahead of end users, and it makes sense to include the electricity sector first in the country's national carbon market.

And hopefully, other variables so far exogenous to the current parameters of our model—e.g., technology breakthroughs, including additional PPA instruments—may come along the way.

[12] As a policy recommendation, one way to consider imposing carbon taxation would be: (a) to figure out the sector-specific average level of carbon intensity, say, in every 3-5 years; and (b) to impose progressive taxes on those that are above the average level of carbon intensity and subsidize those below the average level for technology innovations toward further decarbonization.

[13] In terms of wind power, in light of the great heterogeneity of wind source across the vast territories of China, enhanced regional coordination in stable power supply would be critical.

Challenges and Opportunities

China's goal of achieving carbon neutrality by 2060 means profound transformation of the world's second-largest economy and the biggest emitter. Indeed, just as Chinese President Xi Jinping said at the U.N. General Assembly in September 2020, China must undergo a "green revolution". To achieve carbon neutrality by 2060 is a very tall order, as China currently relies heavily on coal—one of the most carbon-intensive fossil fuels. The other challenge is the relatively short time. The EU plans to spend 50–70 years from carbon emission peak to carbon neutral, but China only has a thirty-year schedule. There are numerous challenges in the nexus of policy, technology, and economy. The time constraint makes the challenges even larger.

But there is also a bright side. Looking ahead, China still enjoys the advantage of backwardness. Its urbanization was about 18% 40 or so years ago, and it has now reached about 60%. By comparison, it is 92% for Japan and 82% for the United States (United Nations, 2018). Moving forward, there is huge room for building smart-cities. Smart-cities integrate real-time data from diverse data sources into a single data hub, where huge amounts of data are analyzed and used to inform management and planning decisions. Instead of retrofitting old cities, smart-city projects can involve building brand-new cities that are eco-friendly and energy efficient from the ground up by deploying low-carbon, zero-carbon and negative-carbon technologies simultaneously. As a top priority for designing and building smart cities in the future, energy efficiency towards carbon neutrality can be optimized through a combination of sensor technology, big data, and data science.

And within our PPA + MBA + CCS + NBS framework, PPA aside, CCS, MBA, and NBS may have higher potential on carbon removal than we had predicted. As we have seen in previous chapters (esp. Chapters 1, 3, 4, 5, and 6), in recent years China has made remarkable progress in CCUS/CCS pilots and has learnt a great deal from its own experience of ETSs as well. As a negative CO_2 emission technology, CCUS, especially CCS + EOR, will be more attractive when the carbon price is sufficiently high and when the technology becomes more mature or advanced.

Expansion of green financing—a typical example of MBA—also helps in terms of financing CCS or CCUS, and we can expect more applications of CCUS years down the road. Generally, green loans, green bonds and green insurance can be powerful tools—via price signaling, risk management and resource allocation—to facilitate industrial transformation, including energy structure from fossil to non-fossil fuels. We recommend that much more work be done on that front for China to achieve carbon neutrality by 2060. Indeed, to phase out coal more aggressively with renewables will be particularly important for China, as our simulation model indicates. There, the good news is that since 2016 when China held the presidency of G20 summit meeting and included the topic of "green finance" into the G20 agenda, the country has made impressive progress in green finance. According to the latest statistics by the Research Bureau of PBOC, by the end of 2020, green loans in China totaled 12 trillion RMB or about 1.8 trillion US dollars (largest in the world), and the stock of green bond reached 813.2 billion RMB or about 125 billion US dollars (second largest in the world, next to the US). Their estimation was that so far the yearly average of green financing in China was equivalent to a reduction of 100 $MtCO_2$ emissions, more than 1% of the national total (9,800 $MtCO_2$ in 2017), and the number is expected to rise years ahead.

There is also good news on NBS, Wang et al. (2020) have recently published a pioneer research in *Nature*. They reevaluate the carbon absorption capacity of terrestrial ecosystem in China and find terrestrial ecosystems, including forests, absorb about 1.11 billion tons of carbon annually from 2010 to 2016, which is 45% of anthropogenic carbon emissions in the same period. While their calculation, i.e. 45%, may have been overestimated and our own estimation is about 10%, their research does suggest that the role of forest carbon sink should be examined more closely. The wood products in China also have great potential (Xu et al., 2020), and the potential for increasing carbon sinks in wood products is estimated at 1.15 billion tons per year around 2060.

Uncertainties

As indicated above, many uncertainties remain around achieving carbon neutral in China by 2060. R&D for de-carbonization strategies seems a least-regret option. Indeed, looking at the endless frontier of science and technology, scientists and engineers around the world have been working on nuclear fusion of hydrogen, and if one day that becomes successful both scientifically and commercially, it will be a game changer in energy transformation. It means almost boundless supplies of zero-CO_2, clean, and safe energy on Earth. But as history shows, technology only evolves and comes with many contingencies. Alternative strategies are necessary when progress of technologies is not on schedule. In transition period, the conflict of interests is unavoidable, particularly in the energy complex.[14] Our model is incapable for the simulation of conflict of interests, the politics of which,

[14] The current dual-track system of ETSs in China—seven regional markets and one national market—is an illustration of the latent issue of conflict of interest unique to the political economy of the vast country. Given different industrial structures across different regions

especially geopolitics, in fact carries high social and economic costs. There are also uncertainties in the deployment of renewable energy in sectors that are more difficult to be electrified in the near future. These sectors are those with high temperature industrial processes, city transportations, trains, shipping, and aviation, etc. In any case, for electrification to work as a tool for decarbonization, the precondition is that the source of power has to be made clean first, that is, in forms of non-fossil rather than fossil such as coal, which dominates China's energy structure today.[15] Otherwise, as a net outcome in the energy complex, one may end up emitting more CO_2 than reducing it in electrification.

A final note: Globally, in the European Union, for instance, of late there has been talks or proposals about transborder carbon taxes. How that will figure in the prospect of Chinese exports (e.g., crude steel or steel products) is an area that needs to be further studied.[16]

---

in China, arguably one national ETS covering all industrial sectors and all at once without transition period runs the risk of widening rather than narrowing the already palpable regional imbalances in terms of both economic development and wealth distribution. Unlike regional ETSs which cover different industrial sectors, China's national ETS, as currently designed, only covers the power sector, and electricity is more or less homogenous across the whole country.

[15] For instance, presently about 60% of power or 1.08 billion kw of electricity in China is coal-fired, and much of it was installed in the past 15 years, and normally won't retire until 20–30 years later. Early retirement will incur additional costs *ceteris paribus*—a daunting task especially for economically poor regions of China.

[16] If the US and EU decide to impose transborder carbon tax, that would have significant impact on Chinese exports, as well as on exports from other developing countries. A critical variable to consider here is to what extent would the transborder carbon tax be earmarked for green technology transfers from the developed to the developing countries.

Acronym	**Full spelling**
BECCS	Bioenergy with carbon capture and storage
CCS	Carbon capture and storage
CCUS	Carbon capture, utilization and storage
DICE	Dynamic integrated model of climate and the economy
EU	European Union
GHG	Greenhouse gas
IUCN	The International Union for the Conservation of Nature
MBA	Market-based approach
NBS	Nature-based solutions
NDC	Nationally determined contributions
PPA	Public policy approach

References

Bellassen, V., & Luyssaert, S. (2014). Carbon sequestration: Managing forests in uncertain times. *Nature, 506*(7487), 153–155. https://doi.org/10.1038/506153a

China Ministry of Environmental Protection. (2019). *China's climate change policy and action 2019 annual report*. Retrieved from http://www.mee.gov.cn/ywgz/ydqhbh/qhbhlf/201911/P020200121308824288893.pdf

Cohen-Shacham, E., Walters, G., Janzen, C., & Maginnis, S. (2016). *Nature-based solutions to address global societal challenges*, xiii+97. IUCN.

Ding, S., Zhang, M., & Song, Y. (2019). Exploring China's carbon emissions peak for different carbon tax scenarios. *Energy Policy, 129*, 1245–1252.

Duan, H., Mo, J., Fan, Y., et al. (2018) Achieving China's energy and climate policy targets in 2030 under multiple uncertainties. *Energy Economics, 70*, 45–60.

Dong, H., Dai, H., Geng, Y., et al. (2017). Exploring impact of carbon tax on China's CO_2 reductions and provincial disparities. *Renewable and Sustainable Energy Reviews, 77*, 596–603.

Gallagher, K. S., Zhang, F., Orvis, R., Rissman, J., & Liu, Q. (2019). Assessing the policy gaps for achieving China's climate targets in the Paris Agreement. *Nature Communications, 10*(1), 1–10.

Hardin, G. (1968, December 13). The tragedy of the commons. *Science, 162*(3859), 1243–1248.

He, G., Lin, J., et al. (2020). Author correction: rapid cost decrease of renewables and storage accelerates the decarbonization of China's power system. *Nature Communications, 11*, 2486.

He, J., Gou Di, Z., Zhan, T. F., Zhuan, Y., Lu, X., & Jiu, J. Y. (2020) [EB/OL]. http://iccsd.tsinghua.edu.cn/wap/news-301.html

He, J., Zhong, G., Di, T., Fa, Z., Yu, Z., Xing, L., & Jing, Y. J., [EB/OL]. (2020). http://iccsd.tsinghua.edu.cn/wap/news-301.html.

Hyman, R. C., Reilly, J. M., Babiker, M. H., & De Masin, A. (2002). Modeling non-CO_2 greenhouse gases. MIT, US.

IEA, 2020. World Energy Outlook 2020. Available from: http://www.worldenergyoutlook.org/.

IPCC. (2018). *Global warming of 1.5°C. An IPCC Special Report on the impacts of global warming of 1.5°C above pre-industrial levels and related global greenhouse gas emission pathways, in the context of strengthening the global response to the threat of climate change.*

Jin, L., Yi, Y., & Xu, J. Forest carbon sequestration and China's potential: The rise of a nature-based solution for climate change mitigation. *China Economic Journal*, 200–222. https://doi.org/10.1080/17538963.2020.1754606.

Li, C., Zhou, C., Peng, W., Lv, Y., & Luo, X. (2020). Accurate prediction of short-term photovoltaic power generation via a noval double-input-rule-modules stacked deep fuzzy method. *Energy, 212*, 1–13.

Lin, B., & Jia, Z. (2019). Energy, economic and environmental impact of government fines in China's carbon trading scheme. *Science of the Total Environment, 667*, 658–670.

Nature. (2017). Nature-based solutions' is the latest green Jargon that means more than you might think. *Nature, 541*(7636), 133–134. https://doi.org/10.1038/541133b

Niu, S., Liu, Y., Ding, Y., et al. (2016). China's energy systems transformation and emissions peak. *Renewable and Sustainable Energy Reviews, 58*, 782–795.

Nordhaus, W. D. (2007). A review of the Stern review on the economics of climate change. *Journal of Economic Literature, 45*(3), 686–702.

Nordhaus, W. D., & Boyer, J. (2000). *Warming the world: Economic models of global warming*. The MIT Press.

Shan, Y., Zhou, Y., Meng, J., Mi, Z., Liu, J., & Guan, D. (2019). Peak cement-related CO_2 emissions and the changes in drivers in China. *Journal of Industrial Ecology, 23*(4), 959–971.

Song, J., Yang, W., & Wang, S. (2018). Exploring potential pathways towards fossil energy-related GHG emission peak prior to 2030 for China: An integrated input-output simulation model. *Journal of Cleaner Production, 178*, 688–702.

Stern, N. (2006). *The economics of climate change: The stern review*. Cambridge: Cambridge University Press.

UNFCCC. (2019). *Climate ambition alliance: Nations renew their push to upscale action by 2020 and achieve net zero CO_2emissions by 2050.*

United Nations. (2018). Department of Economic and Social Affairs, Population Division (2018). World Urbanization Prospects: The 2018 Revision, Online Edition.

Wilkerson, J. T., Leibowicz, B. D., & Turner, D. D. (2015). Comparison of integrated assessment models: Carbon price impacts on U.S. energy. *Energy Policy, 76*, 18–31.

Yuan, J., Xu, Y., Hu, Z., Zhao, C., Xiong, M., & Guo, J. (2014). Peak energy consumption and CO_2 emissions in China. *Energy Policy, 68*, 508–523

Yuan, Y., Duan, H., Tsvetanov, T. G. (2020). Synergizing China's energy and carbon mitigation goals: General equilibrium modeling and policy assessment. *Energy Economics*, 104787.

Xu, J., Da, G., Cheng, N., Yu, Z., Guo, N., Xing, Z. (2020). [EB/OL]. https://mp.weixin.qq.com/s/-dlRSV1b5WyKfn-nEcGfdA

Zhang, P., Shi, X., Sun, Y., Cui, J., & Shao, S. (2019). Have China's provinces achieved their targets of energy intensity reduction? Reassessment based on nighttime lighting data. *Energy Policy, 128*, 276–283.

Correction to: Climate Mitigation and Adaptation in China

Jun Fu, Dongxiao Zhang, and Ming Lei

Correction to: J. Fu et al. (eds.), *Climate Mitigation and Adaptation in China*, https://doi.org/10.1007/978-981-16-4310-1

In the original version of the book, Sect. 4.2 in Chap. 1 was revised. A new sentence has been added on page 29. The chapter and the book have been updated with the change.

In the original version of the book, Chap. 4 was revised. A new sentence has been added on page 109. The chapter and the book have been updated with the change.

The original version of Chap. 5 was updated with few word corrections on page 198. The chapter has been updated with the changes.

In the original version of the book, the following belated corrections were updated: In Chap. 7, a new sentence has been added on pages 254 and 272, respectively, in Sects. 2.1 and 4.4, and the symbol in the Table 2 note has been changed from "$" to "¥" in Sect. 4.1. The chapter and the book have been updated with the changes.

The updated versions of these chapters can be found at
https://doi.org/10.1007/978-981-16-4310-1_1,
https://doi.org/10.1007/978-981-16-4310-1_4,
https://doi.org/10.1007/978-981-16-4310-1_5,
https://doi.org/10.1007/978-981-16-4310-1_7.

J. Fu et al. (eds.), *Climate Mitigation and Adaptation in China*,
https://doi.org/10.1007/978-981-16-4310-1_8

Index

J. Fu et al. (eds.), *Climate Mitigation and Adaptation in China*,
https://doi.org/10.1007/978-981-16-4310-1

M

N

The manufacturer's authorised representative in the EU is Springer Nature Customer Service Centre GmbH, Europaplatz 3, 69115 Heidelberg, Germany. If you have any concerns regarding our products, please contact ProductSafety@springernature.com

Printed and bound by CPI Group (UK) Ltd, Croydon, CR0 4YY
02/10/2026
02227130-0001